Where do natural hazards happen and how do we prepare for them?

OpenSciEd

MIDDLE SCHOOL SCIENCE

Where do natural hazards happen and how do we prepare for them?

Tsunami

OpenSciEd

OpenSciEd

Where do natural hazards happen and how do we prepare for them?

Tsunami

The development of this material was supported by the Bill & Melinda Gates Foundation, Carnegie Corporation of New York, Charles and Lynn Schusterman Family Foundation, and William and Flora Hewlett Foundation.

To download the handouts, slides, and other unit resources, visit OpenSciEd.org and complete the free registration. Accessible versions of the Student Edition, for students with disabilities, are available in the NiMAC Inventory.

TABLE OF CONTENTS

Student Procedures

Student Procedures

Lesson 1: What happens to a community when a tsunami occurs?

Read about a Natural Hazard

On your own

1. Complete Part 1 of your handout individually:
   - Read about an event called a tsunami.
   - Examine photographs from the tsunami.
   - Underline or circle things you notice and wonder about in the reading or photos.

Explore the Tsunami Phenomenon

In your notebook

2. Make a Notice and Wonder chart on a page on the left side of your notebook. Record what you noticed and wondered about from the reading and photos.
3. Watch the videos of the tsunami as a class.
4. Record what you notice and wonder about from the videos in your notebook.
5. Be prepared to share with the class.

Brainstorm Ideas

The goal of this activity is to think of ideas for how we can detect, warn people, and minimize the damage from a tsunami.

On your own

6. Individually, in Part 2 of your handout, brainstorm these ideas for:
   - technologies to help detect a tsunami
   - technologies to give people more advance warning that a tsunami is approaching
   - technologies that could help reduce the damage from a tsunami

7. Look back at your ideas for detecting, warning people, or reducing damage from a tsunami and do the following:
   - Identify 1 idea and describe in detail in Part 2 of your handout.
   - Draw and write about how this technology or design would work to detect, warn people, and/or reduce damage.
   - Select an idea from a different category. Draw and write about how it works.

With a partner

8. Rotate around the room and discuss developed ideas in pairs with 3 different people. Each person will describe one of their developed ideas and why it might work to either
   - detect,
   - warn people, or
   - reduce damage.

9. As you listen to your partners' ideas, write them down in Part 3 of your handout.
10. Be prepared to share with the class.

Share Ideas to Detect, Warn People, or Reduce Damage

Scientists Circle

11. Share an idea that either you or someone else in the class has. Be sure to:
    - describe the idea
    - explain how it could either detect, warn people, or reduce damage

Evaluate Initial Designs

On your own

12. Looking at the class ideas for detecting, warning people, or reducing damage, do these things individually in Part 4 of your handout:
    - Choose the one that you think would be the most promising solution to implement and explain why.
    - Choose the one that you think would be the most challenging solution to implement and explain why.

Scientists Circle

13. Share with the class which solutions you think would be most promising or most challenging. Be prepared to explain why in each case.

Related Phenomena

The purpose of this activity is to share what you know about other natural hazards that might help us think about how different places are at risk for different hazards, and whether the things we know about those hazards might help us design solutions for the tsunami hazard.

With a partner

14. Discuss any natural hazards other than tsunamis that you:
    - have experienced
    - were affected by
    - have heard or read about
    - are interested in learning more about

15. Think in particular about which natural hazards are more likely to affect your community.

On your own

16. Some places are more at risk from certain natural disasters than others. In Part 5 of your handout, answer these questions:
    - Which natural hazard is the area where you live most at risk for?
    - Are there things you know about those hazards that might help us design solutions for the tsunami hazard?

17. Be prepared to share with the class.

What ideas can we get from related phenomena?

With your class

18. Share your related phenomena as a class. As you listen to your classmates' ideas, think about which are examples of:
    - natural hazards that you or someone you know might have experienced
    - natural hazards our community or town is at risk for
    - technologies or other efforts to detect, warn people, or reduce damage from the hazard

Home Learning

19. Talk with someone in your family who is older than you. Tell them you are studying natural hazards.
    - Ask them to share any stories they have about natural hazards.
    - Ask how they remember learning about or preparing for the hazard.

20. Be prepared to share in the next class.

Turn and talk

21. Share the following with your neighbor:
    - What did you learn about natural hazards from your family?
    - What surprised you the most after talking with your family?

With your class

22. Discuss the following with your class:
    - How do you think people determine which places are at risk for a natural hazard?
    - How do you think people compare and evaluate solutions for reducing the impact of a natural hazard?

Driving Question Board (DQB)

On your own

23. Look back at the class's ideas and questions for tsunamis, other natural hazards, and designs for engineering solutions to hazards.
24. Use the class's ideas to generate questions about tsunamis, and natural hazards, specifically about:
    - What still remains unclear?
    - What do we want to know more about?
25. Use the class's ideas to generate questions about engineering design, specifically about:
    - How do we design new solutions to reduce the impact of tsunamis?
    - How do we compare and evaluate engineering designs related to natural hazards?

26. Write at least 1 question related to tsunamis or a other natural hazard and at least 1 question about engineering designs for natural hazards on sticky notes for the DQB.
- Take turns sharing your questions with a partner.
- Be sure to ask clarifying questions when you are unsure of something in a question you hear.

With your class

27. How to build a DQB:

a. The first student reads their question aloud to the class, then posts it on the DQB near the part of the model the question most relates to.
b. Students should raise their hand if any of their questions relate to the question that was just read aloud.
c. The first student selects the next student whose hand is raised.
d. The second student reads their question, says why or how it relates, and posts it near the question it most relates to on the DQB.
e. The student selects the next student to share a related question or a new question.
f. We will continue until everyone has at least 1 question on the DQB.

Ideas for Investigation and Information We Need

With your group

28. Discuss ideas for the best data and information that will help you answer questions on the DQB.
29. Generate a list of ideas with your group.

On your own

30. Record your group's list of ideas in your science notebook on a new page titled, "Ideas for Investigation and Information We Need." Also record the name of your assigned category of questions.

With your class

31. Share out the ideas your group generated.

Lesson 2: Where do tsunamis happen and what causes them?

Revisit the DQB and Data We Need

With your class

1. Revisit the Ideas for Data and Information We Need poster and share why this investigation and data would be helpful.

Making Predictions

With a partner

2. What ideas do you have about where tsunamis happen? Why do you think they happen there?

In your notebook

3. Using *Tsunami Predictions*, circle or color in places where you think tsunamis might occur. Explain why you think they happen in those places.

Be prepared to share with the class.

Where have tsunamis happened in the past?

In your notebook

4. Attach *Where do tsunamis happen?* to your science notebook. Title the page "Past Tsunami Data."

With your class

5. Prepare to analyze the data by looking closely at the information provided:
    - The title and text with the map. What information do these give you? What questions are being asked?
    - The legend of the map. What data is on the map?

6. Analyze the data on https://arcg.is/SDazD . Look for evidence to answer these questions:
    - What patterns do you notice in where tsunamis have occurred?
    - Where do they not occur?
    - What causes the most tsunamis? How do you know?

Be ready to share.

Comparing All Earthquakes to Earthquakes That Cause Tsunamis

With your class

7. Attached *Comparing All Earthquakes to Earthquakes that Cause Tsunamis* to your science notebook. Title the page "Compare All Earthquakes to Earthquakes that Cause Tsunamis." Then, analyze new data using a swipe map at https://arcg.is/1HiOKi . Consider these questions:

- How would you describe the pattern of where tsunami-generating earthquakes occur?
- Are there places in the world that have earthquakes but do not have tsunamis?
- Look closely at the plate boundaries. Do you notice any patterns?

Share Patterns Comparing Earthquakes to Tsunamis

With your class

8. Share an interesting pattern you noticed between all earthquake data and earthquakes that cause tsunamis.

Figuring Out Places at Risk for Tsunamis

With your class

9. How did the data we analyzed help us answer our question, "Where do tsunamis happen and what causes them?"

10. Consider how this data may help us predict what parts of the world are at risk of tsunamis.

Connecting Earthquakes and Tsunamis

We know that most tsunamis are caused by earthquakes. We now know that some kinds of earthquakes cause tsunamis, but not all earthquakes.

On your own

11. What do you think is unique or special about the earthquakes that cause tsunamis? What data would you need to see if you are right?

Digging into the Data

With your group

12. Follow the instructions on *Connecting Earthquakes and Tsunamis* to analyze and interpret the data on the graphs at https://www.openscied.org/tsunami-data-set/ .

13. As you work, consider these questions:

- What is the data set showing?
- How do these data help us (or not) understand what is special or unique about earthquakes that cause tsunamis?

Where do tsunamis happen and what causes them?

With your class

14. What can we now say about these questions:
    - What causes tsunamis?
    - What doesn't cause tsunamis?

In your notebook

15. Revisit your initial predictions map: Which places in the world are most at risk for tsunamis, based on our analyses?
Use another color to show your new predictions for where you think communities are more at risk for tsunamis.

Track Science and Engineering Ideas

With your class

16. Think about the science and engineering ideas you figured out through this investigation. Share your thinking as a class in response to these questions:
    - What science ideas did we figure out?
    - How can engineers use these ideas?

How does a tsunami form because of an earthquake?

With a partner

17. How does a strong, shallow earthquake cause a tsunami to form?

On your own

18. Ask family and friends about what natural hazards have happened where you live. Find out what caused those hazards to occur.

Lesson 3: What causes a tsunami to form and move?

Navigation

With a partner

1. Brainstorm with a partner: How could an earthquake that happens under the ocean result in a tsunami?

Foil Pan Model

With your class

2. Watch the video and make observations about what is happening with the water in the pan. Track your observations on your *Wave Investigations* handout under the "What I see" column.

With a partner

3. Use the "What it means" column to describe how your observations help you understand how a tsunami forms and moves.
Be ready to share with your class.

NOAA Tsunami Model

With your class

4. Watch the computer visualization of the 2011 Japan tsunami and make observations about what is happening as the water moves across the ocean. Track your observations on *Wave Investigations* under the "What I see" column.

With a partner

5. Use the "What it means" column to describe how your observations help you understand how a tsunami forms and moves.
Be ready to share with your class.

Navigation

With a partner

6. Turn and talk with a partner:
    - What have we figured out so far about what happens with the wave(s) when an earthquake results in a tsunami?
    - What do we still need to figure out?

Tsunami Wave Model

With your class

7. Watch the side-view computer simulation and make observations about what is happening from a different perspective. Track your observations on *Wave Investigations* under the "What I see" column.

With a partner

8. Use the "What it means" column to describe how your observations help you understand how a tsunami forms and moves.
Be ready to share with your class.

Comparing Different Types of Models

With a partner

9. Reflect on the three models. Discuss how analyzing different models can help us make sense of how tsunamis are caused by an earthquake.
Be ready to share your ideas.

With your class

10. With your class, identify the benefits and limitations of the different wave models we have analyzed.

Track Science Ideas and Update the Tsunami Chain of Events

Scientists Circle

11. Now that we have analyzed different models or representations of tsunamis, what have we figured out that helps us explain how an earthquake can result in a tsunami?

a. Add ideas to the second column of the Science Ideas chart.
b. Update the Tsunami Chain of Events poster to include those ideas in tracking how a tsunami forms and moves to shore.
c. Revisit the Science Ideas chart to consider how engineers can use these ideas to protect communities from tsunamis.

Navigation

With a partner

12. Consider: If we could get these same key pieces of data about a different natural hazard, do you think it could be used to predict where that hazard will occur and to figure out ways to protect communities?

Lesson 4: How can we forecast where and when tsunamis will happen and which communities are at risk?

Revisit the Tsunami Chain of Events

With a partner

1. Use the Tsunami Chain of Events and *Forecasting Risk to Communities* to predict which areas are at risk for tsunami damage. If a shallow, high-magnitude earthquake occurs at a colliding plate boundary in the Pacific:
    - Where are people and property at risk for damage from tsunamis?
    - What other pieces of information do you want to know to better determine which places are at risk?
    - How will that extra information help you forecast which places are at risk?

Construct a Scientific Explanation

On your own

2. Imagine that a shallow earthquake of 7.6 magnitude occurs along a colliding plate boundary in the Pacific Ocean. Tsunami warning signals are immediately sent out to countries along the Pacific Ocean. Use what you know about tsunamis and how they affect places along coastlines, along with the map on *Explaining and Forecasting Tsunami Risk*, to:
    - Explain whether this type of earthquake could cause a tsunami to form and how it happens.
    - Rate the level of risk for damage and how quickly each of the four different locations will be impacted.
    - Determine which place is most at risk for damage from a tsunami and explain why you think so.
    - Determine which location should be prioritized for resources to protect against future tsunamis.

Revisit the Driving Question Board

As a group

3. Look through the DQB and place sticky dots on the questions that you think we can now answer.
4. Then, look through the questions that have sticky dots.
    - Under what categories did we find questions that we can now answer?

Answer Questions from the DQB

With a partner

5. Answer questions on the DQB by doing the following:
   - Select a question with a sticky dot.
   - Tape it to the upper left-hand corner of a large index card.
   - Use a marker to answer the question on the index card using pictures and/or words.
6. When you and your partner have finished, place the index card with your answered question under the correct category next to the DQB.

Revisit Initial Ideas

On your own

7. Find *Tsunami: Japan 2011* in your science notebook and look back at your ideas for tsunami detection, warning, and reducing damage.
   - Which designs are still promising?
8. Place a sticky dot next to any ideas you think might be worth investing time and resources into.

Share Your Ideas

Turn and talk

9. Turn to a partner and share one idea you marked with a sticky dot in your handout.

Lesson 5: How can we reduce damage from a tsunami wave?

Revisit tsunami protection designs

With your class

1. Look at your design from Lesson 1. Communities in certain coastal areas need protection from this type of natural hazard. Share some ideas with the class.

Learn about the Ryoishi seawall

Ryoishi, Japan is one community impacted by the 2011 tsunami. Ryoishi had a seawall constructed to protect it from strong waves, but the wall failed during the tsunami.

With a partner

2. Read the case study, *What happened in Ryoishi Bay?*, to learn more about Ryoishi. As you learn about the village and listen to the fisherman, consider these questions:
    a. What are some general characteristics of a tsunami?
    b. What characteristics must the 2011 tsunami wave have had in order to cause this damage?
    c. Why would people continue to live in Ryoishi after it was struck by a bad tsunami in the 1930s?
    d. Why would it be important to consider other solutions to reduce or prevent future tsunami damage?

Be ready to share your ideas with the class.

Identifying the problem and goals for a solution

On your own

3. Using *Evaluating Solutions to Protect Communities from Tsunamis*, write your answers to these questions in Part 1 of the handout:
    a. What is the problem we are trying to solve?
    b. What do we need the solution to do?

Be ready to share your ideas with the class.

Learn about potential real-world designs

Watch a video of a coastal engineer explaining common existing tsunami solutions and testing them using a wave tank.

With your class

4. Before watching the video, read the text in Part 3 of the handout and think about these questions:
    a. Who made the video?
    b. Why did they make the video?
    c. Who is the speaker?
    d. Do we think this is a reliable source of information? Why?

With your class

5. Then, consider these questions:
    a. What criteria are we using to evaluate these solutions?
    b. How will we know if one solution is more effective than another?
6. Analyze the video and record notes in the table in Part 3 of the handout.
7. When engineers evaluate existing solutions, it is important to share findings with everyone involved. Communicate the data from the video in a way that is understandable for others who have not been a part of the data collection process.
    a. How could we rate these solutions in a way that clearly communicates how well they meet the criteria?
    b. What rating systems do other individuals and companies use that can be easily understood and shared with others?

Exit Ticket

On your own

8. Answer the following question on a notecard, or in your science notebook:
    - Is keeping water from reaching the community the only criteria we need to consider? Is there anything else we need to consider when determining which solution would work best for Ryoishi?
        - If yes, explain why.
        - If no, list other ideas for consideration.

Identify Ryoishi's constraints

Turn and talk

9. Talk with a partner about the following question:
    - If we only consider our own criteria when evaluating solutions, will we be picking the best overall solution for Ryoishi? Why or why not?

Determining the needs of the community

With a partner

10. Read *Comparing Ryoishi to Nearby Communities* to learn more about Ryoishi and the surrounding area.
11. Use this information to determine what else we might need to consider when determining the best solution for Ryoishi. List these constraints and why they are important in Part 4 of your handout.

Read about potential solutions

With a partner

12. Read *Existing Solutions for Coastal Communities* about 8 potential existing solutions for the community.
    a. Use the rating scale to learn how well each solution meets the criteria and constraints.
    b. The higher the rating, the better that solution is at meeting the given criteria or constraint.

13. As you read, consider these questions with a partner:
- What are the key features of each type of solution?
    a. How well does the solution meet our criteria?
    b. How well does the solution perform when considering other constraints?

Be ready to share your ideas with the class.

Compare design solutions

With your group

14. Sort the solution ranking cards in order of how well the solutions meet each criteria and constraint.
- Order the solutions from "best for the community" to "worst for the community."

Evaluate designs using a decision matrix

With your group

15. A decision matrix, *Decision Matrix*, will help us better evaluate the existing solutions based on our criteria and constraints. This tool will help determine a good solution for Ryoishi.
    a. Add descriptive labels for our identified criterion and constraints to the green boxes at the top of the matrix.
    b. List each solution's ratings in the corresponding columns next to the solution.
    c. The higher the rating, the better it meets the criteria or constraints.
    d. When you are finished filling in the matrix, try to determine which solution might best meet Ryoishi's needs.

Take stock of ideas

With your group

16. Work with your group to determine which existing solution will best meet the criteria and constraints of the community of Ryoishi.
17. Decide with your group what is important in a solution, and base your decisions on your priorities.

With your class

18. On the class chart, place a sticky dot next to the solution you think will work best for Ryoishi.
19. As you place your sticky dot, consider this question: Why do you think that solution will work better than others?
Be ready to share your reasoning with the class.

Consensus Discussion

Turn and talk

20. Turn and talk to a partner about the following questions:
    a. Looking at your matrix, which solution did you or your group think would work best for Ryoishi?
    b. Why do you think that is the best solution?

Be ready to share your ideas with the class.

With a partner

21. Reread *Comparing Ryoishi to Nearby Communities* and analyze what might be the most important criteria and constraints for Ryoishi.
22. Be ready to share what you think is important for the community and area with the class.
23. Reevaluate your solutions using the primary and secondary criteria.
    a. Work with a partner to reevaluate the designs.
    b. What criteria and constraints did we consider to be more important than others?
    c. Which design solution do you now believe is the best for the community?

Be ready to share your ideas with the class.

Considering solutions in our Tsunami Chain of Events

With your class

24. Let's revisit our Tsunami Chain of Events poster and update it with what we can do as engineers.

Determine next steps

We looked at 8 different potential tsunami solutions for Ryoishi. Each solution seemed like it would work, but both the seawall and breakwater failed to fully protect the towns in past tsunamis.

With a partner

25. Discuss the following prompts with a partner:
    a. If a solution fails, like if a seawall breaks, what else can be done to protect the people in the community from the tsunami?
    b. Are there other design features we brainstormed from the tsunami phenomenon in Lesson 1 that we could look into?

Lesson 6: How are tsunamis detected and warning signals sent?

Navigation

We've looked closely at Ryoishi, a coastal town in Japan that was affected by the 2011 tsunami.

Turn and Talk

1. What happened in Ryoishi during the tsunami?
2. If people want to stay in Ryoishi, what do we need to think about to prepare for future tsunamis?

Close Reading: How do we detect tsunamis?

With a partner

3. Do a close reading of *Reading: How are tsunamis detected and warning signals sent?* with a partner. Discuss and respond to the following questions:
    - What do people need to know in order to prepare for a tsunami?
    - What must a tsunami detection and warning system be able to do?
    - How can we make a diagram of a tsunami detection and warning system working together?
4. Be ready to share with the class.

Update Tsunami Chain of Events poster

With your class

5. Discuss the following prompts:
    - Looking back at the Tsunami Chain of Events poster, at which points do you think a tsunami detection and warning system might play an important role?
    - Are there certain parts of the tsunami detection and warning system that you think connect to more than one part of this diagram?
6. Share your ideas to add to the chain of events.

Navigation

With your class

7. An earthquake has occurred and the tsunami detection system has signaled that a tsunami is approaching. The local government is about to send a warning to people nearby.
    - As a class, discuss what we should be concerned with now.

Lesson 7: What are ways we can communicate with people before and during a tsunami?

What does a tsunami warning signal sound like?

With your Class

1. What do you Notice and Wonder about as you hear these warnings?
2. Share your thinking about what you heard and what you wonder about.

Voices from the Japan Tsunami

With a partner

3. Read at least 3 stories from people who experienced the 2011 tsunami using *Voices from Japan Tsunami Survivors*. Think about the following questions:
    - Who is this person?
    - What was their experience like?
    - What else would you like to know about their experience?

Identify Stakeholders

With a partner

4. Using Part 1 of *Community Stakeholders*, consider the following:
    - Who are the people who live in places where tsunamis might happen?
    - Will they need special assistance in an emergency? Describe those needs.
    - Who can assist them?

Track your ideas on your handout and be ready to share.

Developing Communication Systems

With a partner

5. Using Part 2 of *Community Stakeholders*, answer the following:
    a. What is the problem that engineers are trying to solve when they develop hazard communication systems?
    b. What would that system need to do to address all the needs of the community stakeholders?

Develop Potential Criteria and Constraints

With you class

6. Identify one stakeholder group, one criteria, and one constraint for a communication system.

With a partner

7. Together, brainstorm additional criteria and constraints for a communication system.
   a. Identify two more stakeholder groups.
   b. Then, develop two more criteria and two more constraints for a communication system.
   c. Use these questions to inform your thinking:
      - What media options should be available?
      - Can all people hear, see, or feel some kind of warning signal or alert?
      - What languages or symbols should be used in this kind of system?
      - What does the message need to say to warn people?
      - What are the evacuation routes and places for people to shelter? How are these routes marked?
      - How will people know what to do when the warning signal goes off?

Communication in our World

Home learning

8. Between now and next time we meet, observe all the different ways we send and receive communication with each other. Could we include any of those ideas as well?

Case studies: Communication Options

With your group

9. As you read and view the materials in *Tsunami Communication Examples*, consider:
   - How well does the option fit the criteria? Choose a rating for each criteria.
   - What are the constraints with using this option? List those on the handout.
10. Mark your ratings on *Evaluation Matrix*.

Come to Consensus

With your group

11. Meet with the other group that evaluated the same communication option.
12. Compare ratings to criteria and come to consensus on the ratings.
13. Compare the list of constraints and come to consensus.

Choosing Options

Some communication options do well on some criteria, but not others.

With your class

14. Why would it be important to have multiple modes of communication in the event of a hazard?
15. Would you know what to do when you hear a warning signal?

Case Study: Kamaishi Junior High School

One your own

16. Use the Close Reading Strategy while reading *Reading: Kamaishi East Junior High School.*

- Identify the question(s) you are trying to answer in the reading.
    a. What communication plans did this school have in place?
    b. What worked well or did not work well?
    c. Is there anything about this case that we have not thought of yet?
- Read once for understanding to see what the reading is about.
- Read a second time to highlight a few key ideas that help answer the questions you had.
- Summarize the key idea(s) in your own words, in diagrams, or both.
- Jot down new questions that this raises for you.

With your class

17. Discuss the following questions with your class:
    - What communication plans did this school have in place?
    - What worked well or did not work well?
    - Is there anything about this case that we have not thought of yet?

Update Tsunami Chain of Events

With your class

18. Update the Tsunami Chain of Events with new ideas and how they can be used to protect communities.

Exit Ticket

On your own

19. What ideas do you have for systems the school or community could put in place to make sure students get alerted even when one system breaks?
    - How would we know those systems would also work in an emergency?
    - If we compare each method, do they have the same levels of reliability?
    - Are some systems better than others during an emergency?

Lesson 8: Which emergency communication systems are the most reliable in a hazard?

Revisiting Communication Systems

With your class

1. What were some of the ideas you had last class about which systems work better or are more reliable in an emergency?
2. What does it mean for a system to be reliable?

Gathering Information about Signals

With your group

3. Use a close reading protocol to gather information about warning signals. Use *Obtaining Information Notetaking Guide* to take notes.
4. Access either the multimedia version of the reading at https://arcg.is/1GymWO or mark up a printed copy of *Sending Warning Signals*.

Close Reading Protocol

1. Identify the question(s) you are trying to answer in the reading.
    a. Which emergency communication systems are the most reliable in a hazard?
    b. What makes a system reliable?
2. Read once for understanding to see what the text and media is about.
3. Read a second time to highlight a few key ideas that help answer the questions you had.
4. Summarize the key idea(s) in your own words, in diagrams, or both.
5. Jot down new questions that this raises for you.

Consensus Discussion

With your class

5. What new ideas do we have about warning signals and their reliability?
    - What type(s) of signals should we use during a natural hazard?
    - Should we use only one method, analog or digital? Why or why not?

Exit Ticket

On your own

6. Think about the towns of Ryoishi and Kamaishi, and other coastal towns of Japan. They had seawalls, breakwaters, sirens, alerts, and disaster education at schools. How do all these systems work together to protect communities from tsunamis?

Lesson 9: How can we model the systems put into place to protect communities?

Look back at ways to protect people and property

With a partner

1. Create a chart on the left-hand side of your notebooks and generate a list of all the design solutions that fit into each of these categories:

Detect and Forecast	Warn and Communicate	Reduce Damage

With your class

2. Share ideas with your class and create a shared chart of all the different design solutions that can protect people and property from tsunamis.

Developing a Tsunami Protection System

Scientists Circle

3. Map the different design solutions to different parts of the Tsunami Chain of Events.
    - Start with the earthquake occurring and the process of a tsunami starting.
    - Then work through the chain of events, adding different design solutions at each part of a tsunami that are designed to detect, warn people, and reduce damage.

Turn and talk

4. Turn and Talk to your partner about the following questions:
    a. Why does this system have so many different components and interactions?
    b. What would happen if a component of a subsystem was removed? Would there be effects on the overall system? Why or why not?
    c. What trade-offs do different communities make in this system?

Be ready to share your ideas with the class.

Conduct the Engineering Self-Assessment

On your own

5. Use *Engineering Self-Assessment* to self-assess how well you did in each area of engineering design during this unit.
    a. Circle the number that best represents how you feel you have performed in that area.
        3 - I did great!
        2 - I did OK.
        1 - I need to work on this, or I did not do this step.

Turn in to your teacher for feedback.

Look Back at Related Phenomena

Turn and talk

6. Look back at the Related Phenomena chart from Lesson 1 and discuss the following questions:
    a. What other natural hazards might affect our community?
    b. What systems are in place like the tsunami system?
    c. What can we do to help our community prepare for a natural hazard?

Be ready to share your ideas with the class.

Lesson 10: How can we effectively prepare our communities for a natural hazard?

Navigation

With a partner

1. Turn and Talk to a partner about the following questions:
   - What natural hazards can affect our community?
   - How do we know our levels of risk for these hazards?

   Be ready to share your ideas with the class.

Data to Assess Risk

With your class

2. Discuss the following ideas with your class:
   - What data did we use to determine which places in the world were at risk for tsunamis?
   - How could we use a similar type of data to assess our risk for other natural hazards?

With your group

3. Look at each natural hazard in the map series. Do the following:
   a. Identify and describe the general patterns of risk for the natural hazard.
   b. Identify the local level of risk for the hazard.
   c. Note questions or curiosities you have about this natural hazard.

Determine Your Risk

Writing in Science

4. Once you have identified and described general patterns of risk and your local risk for each hazard, individually consider and respond to the following:
   - Explain which hazard you are at most risk for in your local community. Use evidence from the maps to support your explanation. If there are additional sources of data that you would want to study, also include those ideas in your explanation.

Exit Ticket

With your class

5. Share your ideas for natural hazards you want to investigate.
6. Generate a list of information or data you need to further investigate these natural hazards.

Determine What to Do With Our Information

With your class

7. Look back at the chart we developed at the end of our last class period.
    - Do you have any additional data or information that you want about your hazard that we should add to our chart?

Turn and talk

8. Consider how to use our information:
    - What should we do with this information about our hazards once it is collected?
    - What do scientists and engineers do with their hazard information?

9. Consider who to communicate with:
    - Who in a community might need to know about hazards?
    - Should we also consider any people outside of the community?
    - Are there any specific groups of people that are in every community at any given time?

Be ready to share your ideas with the class.

Consider Who We Should Inform

Turn and talk

10. Discuss and write down ideas for the following questions:
    - Who in a community might need to know about hazards?
    - Should we also consider any people outside of the community?
    - Are there any specific groups of people that are in every community at any given time?
    - Do all people need to know the same information about a hazard?
    - Are there specific instructions or considerations for different groups of stakeholders?
    - Do all stakeholders use the same methods for getting information?

Be ready to share your ideas with the class.

Determine Hazard Information for Community Needs

With your class

11. Review the information needed to develop a communication plan:
    - About the ____ hazard: An overview of the natural hazard and what kind of damage the hazard can cause.
    - Forecast and Detect: Information about how the hazards are/can be forecasted and detected.
    - Warn and Communicate: How the community knows a hazard is happening and how information is shared with them.
    - Reduce Damage: What can be done before the hazard by members of the community to reduce damage from the hazard.
    - Helpful Resources: Additional sites where you can gather more information about the hazard.

With your group

12. Create your communication plan and final product. Use all the resources available to you and expand to other resources you need.

Revisit our DQB

With your class

13. Evaluate and celebrate progress on the DQB:
    - Place your sticky dots on the questions that match questions you think we have made progress on.
    - Once you have placed your 5 dots, step back to the Scientists Circle.
    - As you are waiting for others to place their sticky dots:
        - Be prepared to share out your ideas for the answers to the questions, along with your evidence for these answers.
        - Notice which questions have the most sticky dots and wonder about why these questions have the most dots.

14. With your class, discuss the following questions:
    - Which questions have we made the most progress on?
    - What have we figured out?

References

Tsunami Images

Kyodo News. Rights Reserved.

Kyodo News. Rights Reserved.

Japan Coast Guard

Japan Coast Guard

Tuva Graphing Instructions

Step 1: Observe the initial setup for the tsunami data set on Tuva.

Navigate to https://www.openscied.org/tsunami-data-set/ . Discuss with your partner(s):

- What data is available on the left menu?
- What do you think the dots represent on the graph?
- How many tsunami cases are available in this dataset?

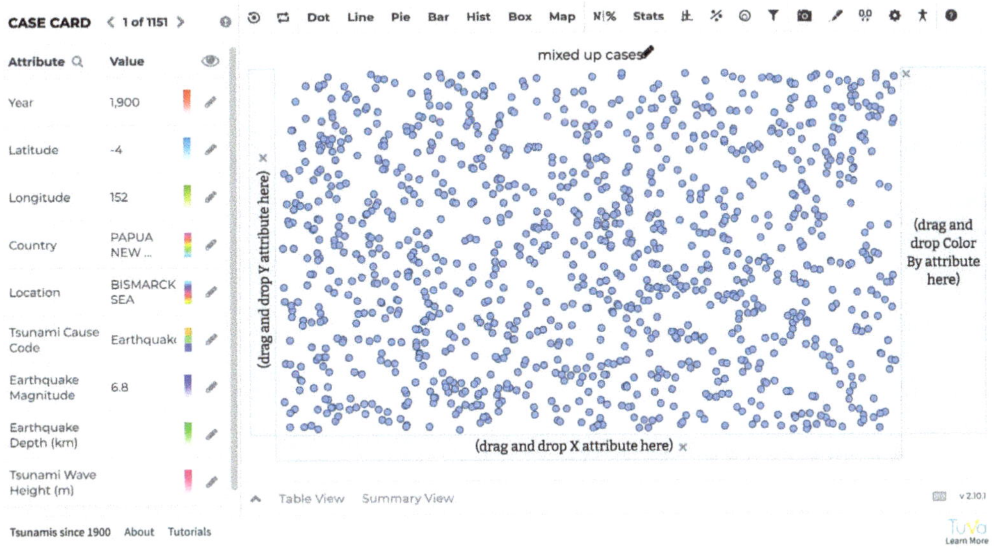

Step 2: Create a graph to represent how the magnitude of an earthquake affects the wave height of a tsunami.

- Click on "Magnitude" from the data menu on the left side of the graph
- Drag it to the *x*-axis box label "Drag and drop X attribute here"
- Click on the "Maximum Water Height" from the data menu on the left side of the graph
- Drag it to the *y*-axis box label "Drag and drop Y attribute here"

Now use *Connecting Earthquakes and Tsunamis* to record the pattern you notice on the graph and what you think it means for how the magnitude of an earthquake affects tsunami wave heights.

To clear the data from the graph, click the "X" next to the "Magnitude" on the *x*-axis and the "Maximum Water Height" on the *y*-axis.

Step 3: Create a graph to represent how the depth of an earthquake affects the wave height of a tsunami.

- Click on "Focal Depth" from the data menu on the left side of the graph
- Drag it to the *x*-axis box label "Drag and drop X attribute here"
- Click on the "Maximum Water Height" from the data menu on the left side of the graph
- Drag it to the *y*-axis box label "Drag and drop Y attribute here"

Now use *Connecting Earthquakes and Tsunamis* to record the pattern you notice on the graph and what you think it means for how the depth of an earthquake affects tsunami wave heights.

To clear the data from the graph, click the "X" next to the "Focal Depth" on the *x*-axis and the "Maximum Water Height" on the *y*-axis.

Historical Data of Tsunamis and Related Events

Locations of historical tsunamis for the past 100 years

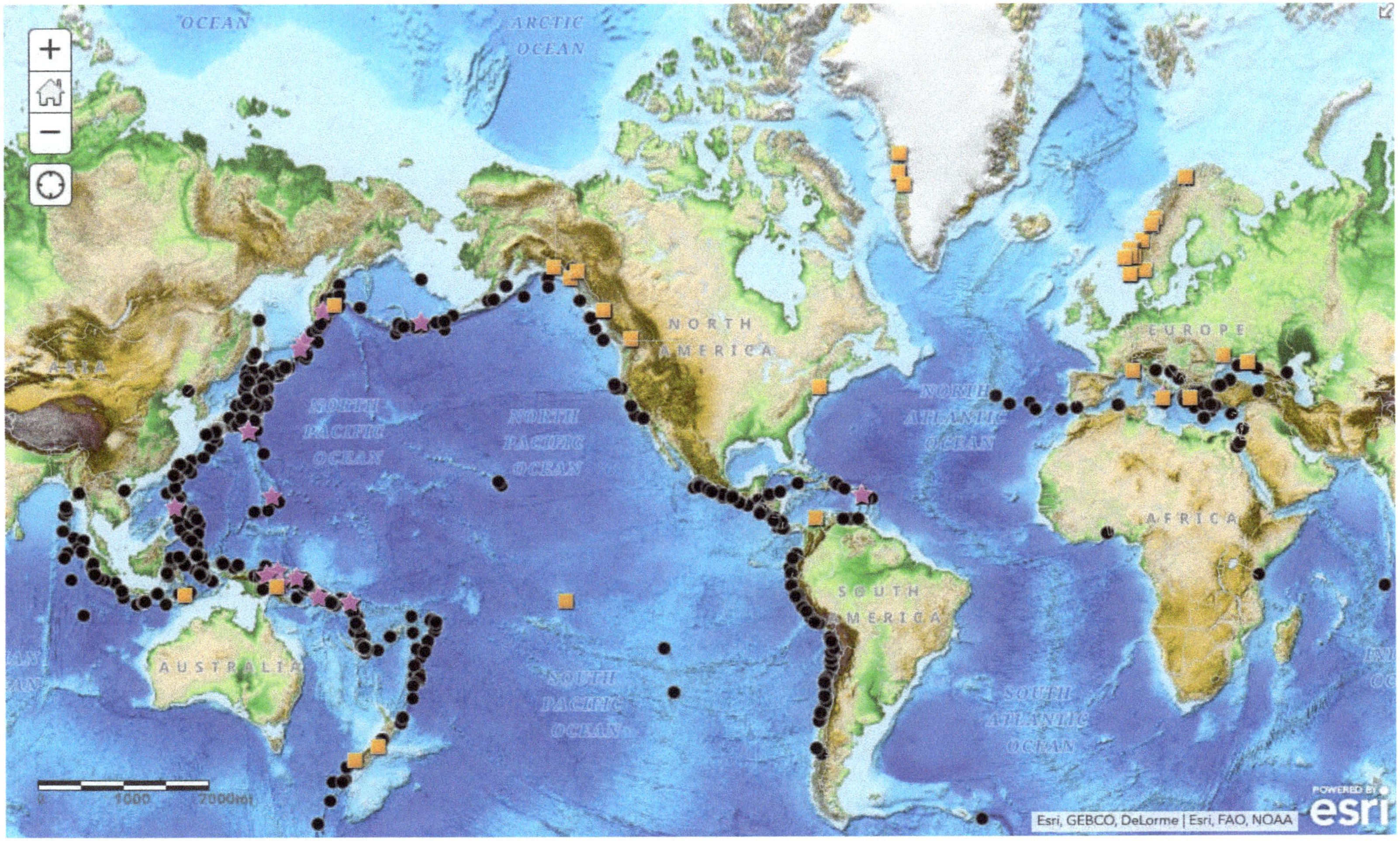

Legend

Cause of Historical Tsunamis

- Earthquake
- Landslide
- Volcano

Tsunamis caused by earthquakes

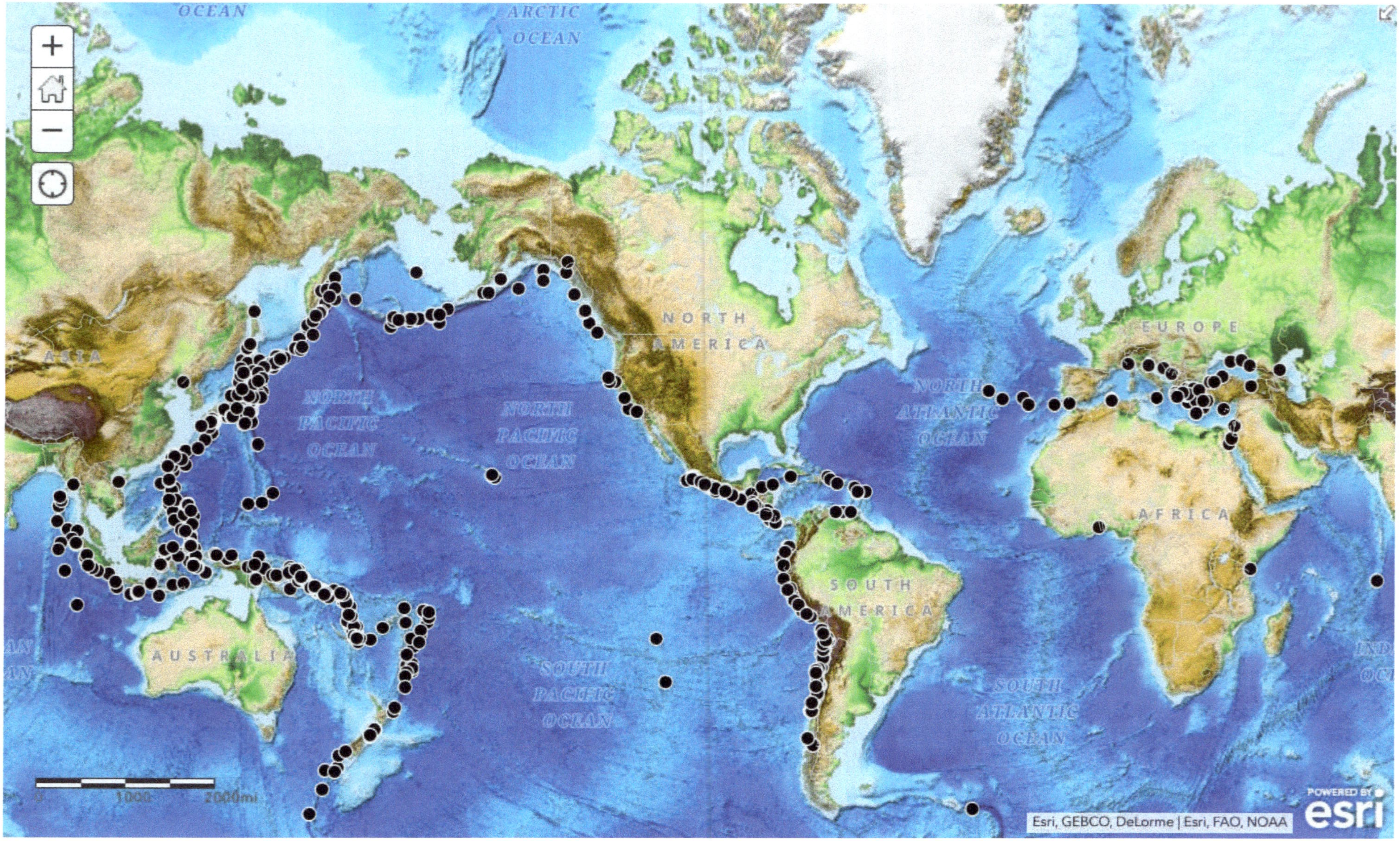

Tsunamis caused by volcanoes

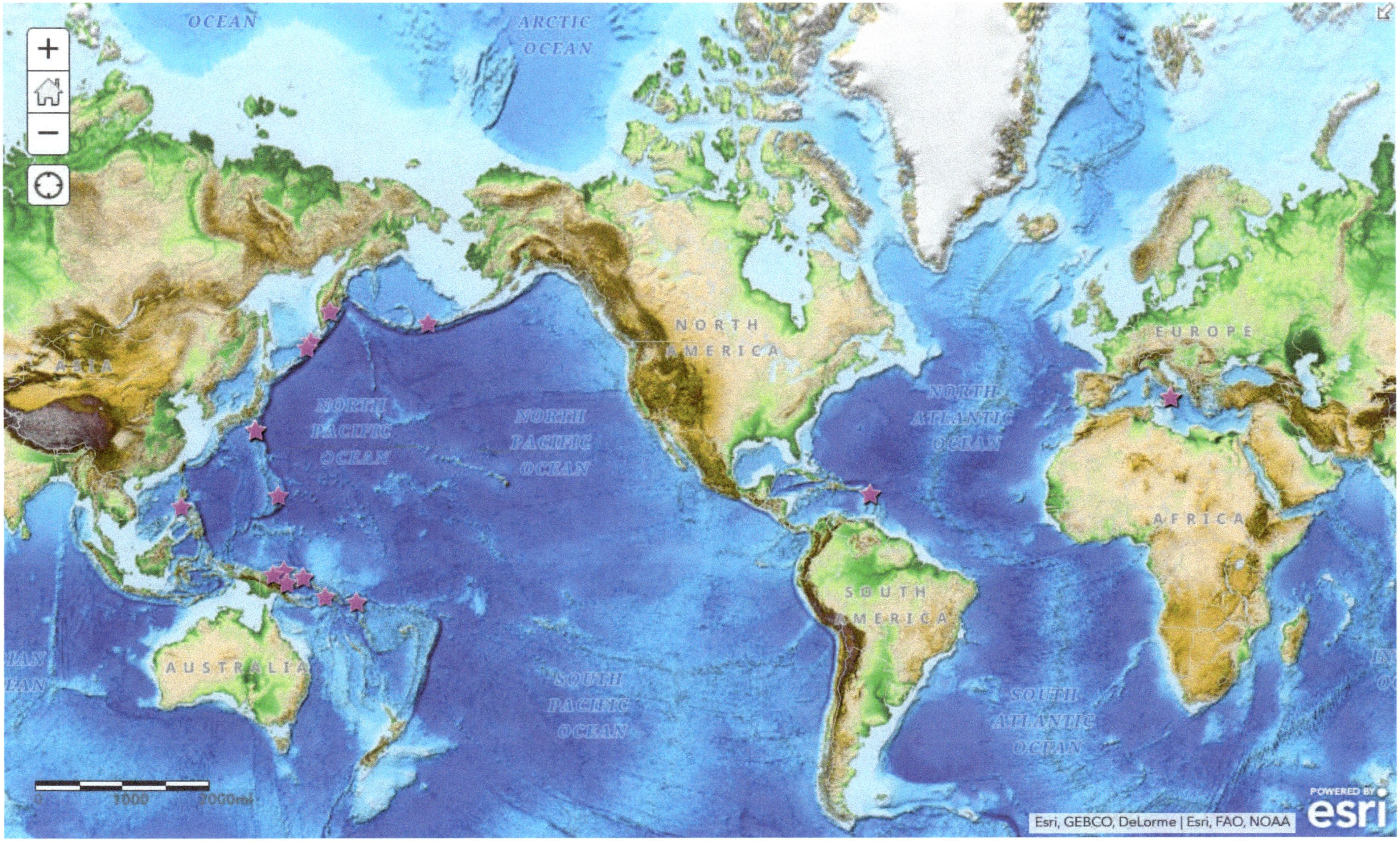

Tsunamis caused by landslides

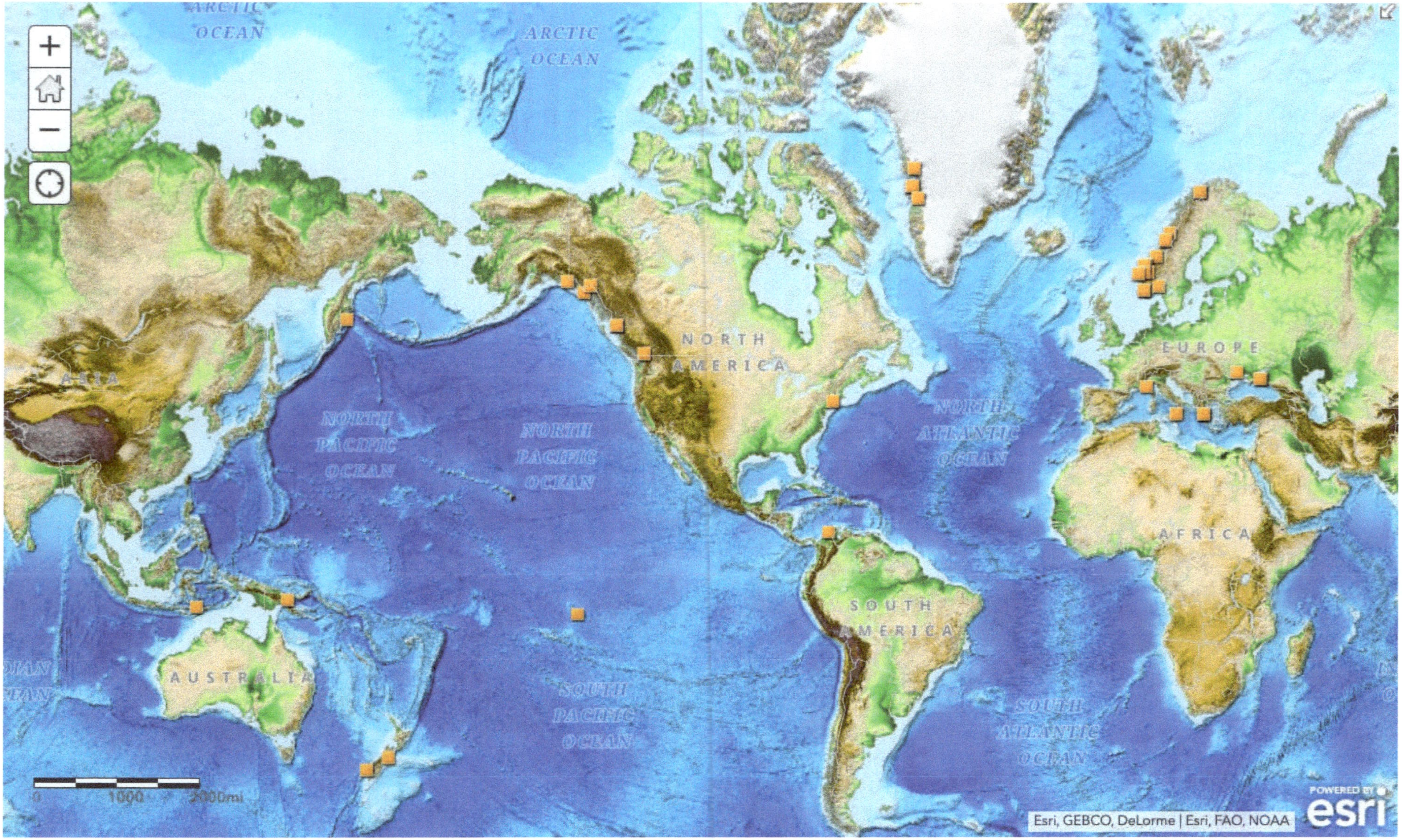

Locations of all recent earthquakes

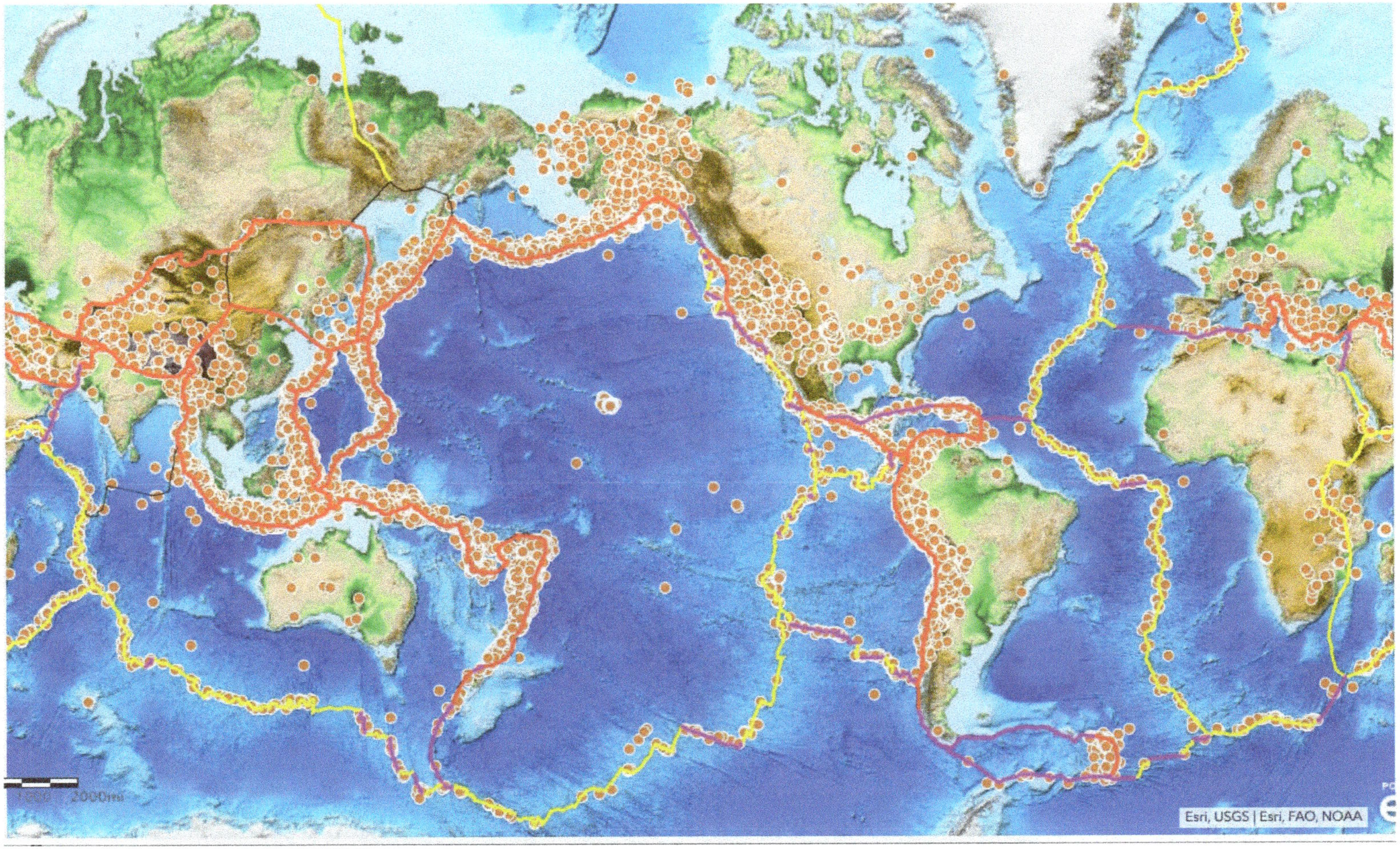

Type of Plate Boundary

- Colliding
- Spreading Apart
- Sliding alongside
- Unknown

Locations of earthquakes that cause tsunamis

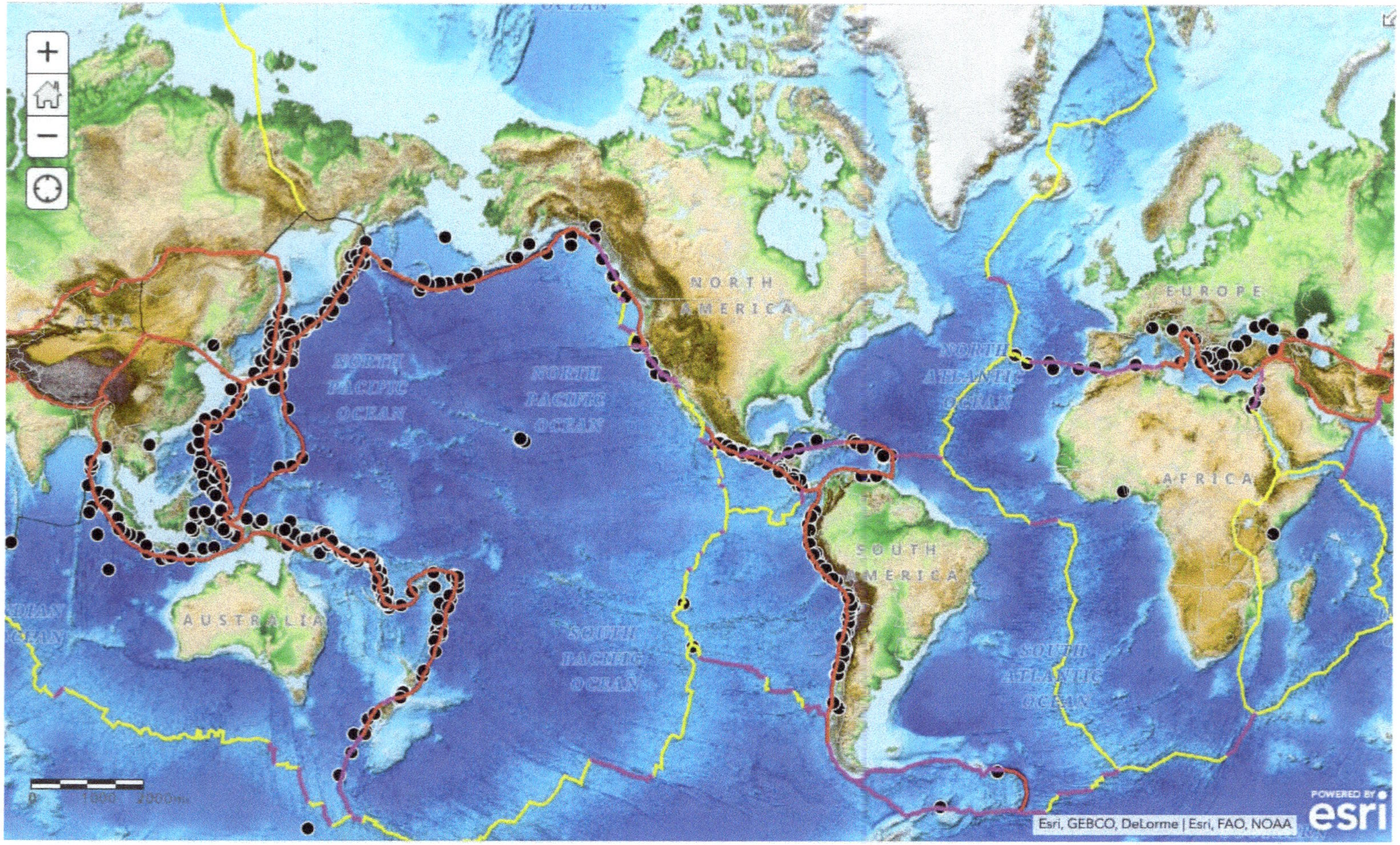

Type of Plate Boundary

- Colliding
- Spreading Apart
- Sliding alongside
- Unknown

Earthquake magnitude versus wave height of the tsunami that forms

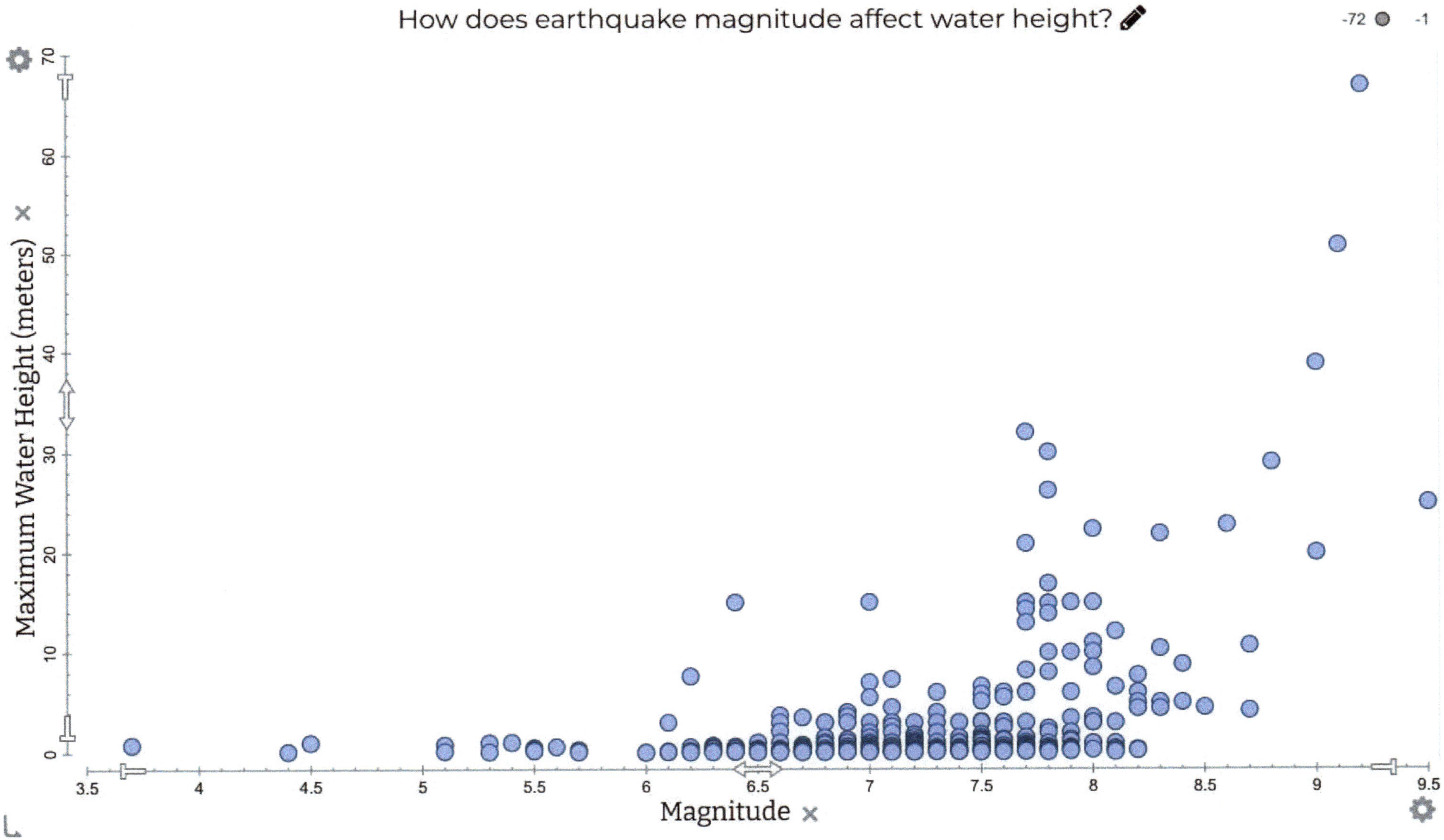

Earthquake depth versus wave height of the tsunami that forms

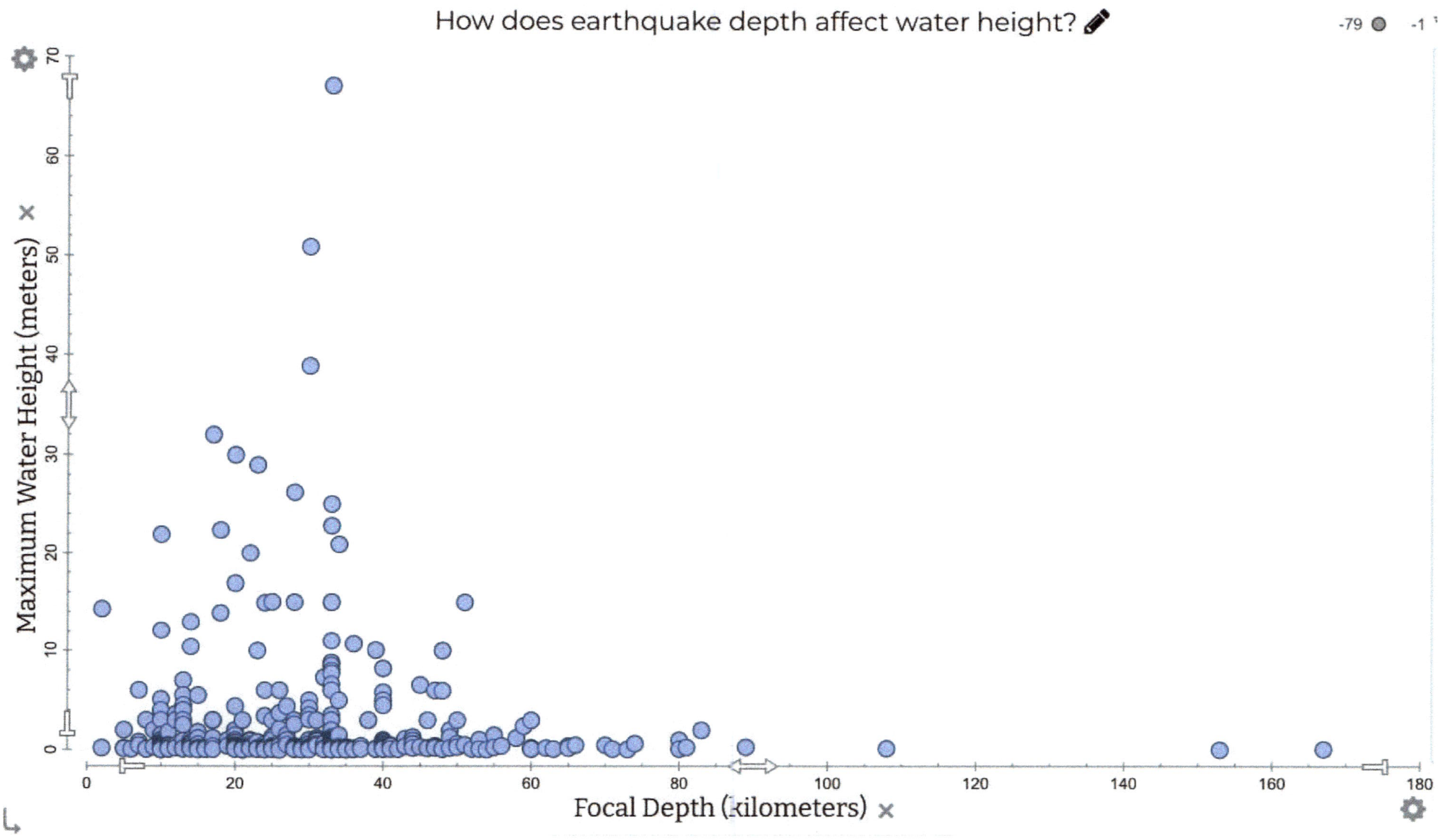

Existing Solutions for Coastal Communities

Proposed Solution # 1: Walls

The first category of solutions advertised to reduce the damage from incoming ocean waves is to build walls. Engineers have designed several types of walls that are effective options to block the flow of water from a tsunami by absorbing some of the wave's energy and also reflecting that energy back out to the ocean.

An example of a vertical seawall.

These walls are typically made of a very hard material, like concrete, and are built high enough to block large waves. The *seawall* at Ryoishi was 30 feet high--some others are over 40 feet. Many walls are completely flat, vertical structures. Other types include *levees* or *sea dikes* as well as *recurved walls. Levees* or *sea dikes* are made of earth or concrete, and they use angled walls or mounds to hold back the water. A *recurved wall* is made of a hard material, like concrete, and uses a curved design to capture and reflect water from waves back to the ocean or sea.

Study the images of each wall type closely and use the data tables on the next page to review how well they perform.

An example of a levee (left) and a recurved wall (right).

Performance of a Seawall

A. SEAWALL		
Criteria & Constraints	**Score**	**Notes**
Ability to break or block waves	★★★☆☆	Can break apart with the force of a large tsunami. Water can flow over it.
Impact on boats/ocean traffic	★★★★★	Built on land, so it does not affect boat traffic.
Impact on marine life	★★★★☆	Marine life is not typically impacted, but creatures that need access to land at some point (like sea turtles) may not be able to reach their nesting grounds.
Impact on ocean view	★☆☆☆☆	In some places, can completely block ocean views.
Cost	★★★☆☆	Expensive to build, but the materials are affordable.
Time to build	★★★☆☆	Large walls can take years to build, but smaller walls can be put into place more quickly.
Maintenance	★★★☆☆	Can last decades, but requires ongoing maintenance to ensure proper functioning.

Performance of a Levee or Sea Dike

B. LEVEE OR SEA DIKE		
Criteria & Constraints	**Score**	**Notes**
Ability to break or block waves	★★☆☆☆	Best for smaller amounts of water, but perform worse when a large wave hits it.
Impact on boats/ocean traffic	★★★★★	Built on land, so it does not affect boat traffic.
Impact on marine life	★★★☆☆	Can impact the flow of water along the shore in coastal ecosystems and affect marine life.
Impact on ocean view	★★★★☆	Smaller than other solutions and usually does not block ocean views.
Cost	★★★★★	Affordable and typically made of rocks, soil, and grass.
Time to build	★★★★☆	Smaller than other solutions and can use local materials, so building it can be quicker.
Maintenance	★★★★☆	Relatively easy to maintain.

Performance of a Recurved Wall

C. RECURVED WALL		
Criteria & Constraints	**Score**	**Notes**
Ability to break or block waves	★★★★★	Can completely absorb and reflect large waves.
Impact on boats/ocean traffic	★★★★★	Built on land, so it does not affect boat traffic.
Impact on marine life	★★★★☆	Marine life is not typically impacted, but creatures that need access to land at some point (like sea turtles) may not be able to reach their nesting grounds.
Impact on ocean view	★★☆☆☆	Can be smaller than a vertical wall because it is curved, so it blocks ocean views less.
Cost	★★☆☆☆	Even more expensive than the other types of walls.
Time to build	★★★☆☆	Requires additional work and time to mold the concrete. A large recurved wall can take years to build, but a smaller one can be put in place more quickly.
Maintenance	★★★☆☆	Can last decades, but requires ongoing maintenance to ensure proper functioning.

Proposed Solution #2: Breakwaters

An example of tetrapods. Hans

The second category of solutions advertised to reduce wave damage is to build *breakwaters*, or structures in the ocean or along the shore that are designed to break up a tsunami wave, absorb some of the wave's energy, and reflect the energy in all different directions, thus decreasing the wave's amplitude as it approaches the shore.

Breakwaters can be found underwater or above water, typically out in shallow parts of the ocean near the shore but also closer to the shoreline as well. Examples of breakwaters include *tetrapods*, *rock armor*, and *submerged rock walls*. *Tetrapods* are concrete structures that reduce a wave's amplitude by absorbing some of its energy while also letting some of the water flow through them. Tetrapods have been used alone but are often found in front of a seawall to help it absorb and reflect energy from waves. *Rock armor* is often found along the shoreline. As with tetrapods, a wave flows through the rock armor and begins to break apart as the water passes between the rocks and reflects off them. A final breakwater example is a *submerged rock wall* located offshore, designed to absorb and reflect the wave's energy before it reaches shore.

Study the images of each type of breakwater closely and use the data tables on the next page to review how well they perform.

An example of rock armor (left) and a submerged breakwater (right).

Performance of a Tetrapod

D. TETRAPODS		
Criteria & Constraints	**Score**	**Notes**
Ability to break or block waves	★★★★☆	Reduces the wave's energy by reflecting it in all different directions. Flooding can still occur as water flows through it.
Impact on boats/ocean traffic	★★★☆☆	Often above the surface, so boats must steer around it in small bays and along the shoreline.
Impact on marine life	★★☆☆☆	Shown to decrease fish populations and disrupt normal erosion along the shoreline.
Impact on ocean view	★★★★☆	Typically does not obstruct ocean views, but can be seen in the horizon.
Cost	★★★★☆	Concrete and molds are shipped to the building site to save costs.
Time to build	★★★★☆	Can be made and stacked quickly onsite.
Maintenance	★★★☆☆	Its sturdy, heavy concrete is more resistant than rock to being washed away, but it does erode from ocean currents.

Performance of Rock Armor

E. ROCK ARMOR		
Criteria & Constraints	**Score**	**Notes**
Ability to break or block waves	★★★☆☆	Can break up waves, but water can still pass through, resulting in flooding. The energy of the wave will be less than if there were no barrier at all.
Impact on boats/ocean traffic	★★★☆☆	Typically found along the shoreline, but can also be built into bays as walls. Boat traffic must navigate around it.
Impact on marine life	★★★☆☆	Provides habitat for some marine life, but may cause loss of habitat for other marine life.
Impact on ocean view	★★★★☆	Not very nice to view on land, but it typically does not obstruct ocean views.
Cost	★★★★★	Rock is an affordable material.
Time to build	★★★★☆	Can be built fairly quickly if the land is owned by the community where it needs to be built.
Maintenance	★★★☆☆	Rock wears down over time.

Performance of a Submerged Breakwater

F. SUBMERGED BREAKWATER		
Criteria & Constraints	**Score**	**Notes**
Ability to break or block waves	★★★★☆	Effective at breaking up a wave, but it can trap water as it is reflected from the shore, which increases the water level near shore.
Impact on boats/ocean traffic	★★★★☆	Only larger, deeper-hulled boats are affected by submerged breakwater.
Impact on marine life	★★☆☆☆	Can influence the flow of water and the movement of marine life in the area.
Impact on ocean view	★★★★★	Underwater, so ocean views are not affected.
Cost	★☆☆☆☆	Can be very expensive to build, depending on the water's depth and the wall's length.
Time to build	★★☆☆☆	Requires heavy, specialized machinery and multiple layers of material, so it can take a significant amount of time to construct.
Maintenance	★★★★☆	When built well, it can last decades, but will slowly wear down due to wave action.

Proposed Solution #3: Natural Vegetation

The third category of solutions advertised to reduce wave damage is to use *natural coastal vegetation* (that is, plants that commonly grow along coastlines) to absorb and reflect the energy from a tsunami wave, similar to breakwaters. They have the added benefit of providing something for people to hold onto or climb for protection during a tsunami event.

An example of mangroves.

Natural vegetation grows in particular climates that provide the right temperatures and rainfall. For example, *mangrove forests* typically grow in tropical and subtropical climates. They need warmer temperatures to grow, so in Japan they are only found in southern parts of the country. *Pine forests* grow in colder climates but take many years to fully mature, making them a less desirable option for communities that need immediate tsunami protection.

An example of a pine forest Karl Hörnfeldt

Performance of a Mangrove Forest

G. MANGROVE FOREST		
Criteria & Constraints	**Score**	**Notes**
Ability to break or block waves	★★★☆☆	Can block much of the energy of a smaller tsunami wave, but is less effective in larger tsunamis.
Impact on boats/ocean traffic	★★★★★	Does not impact normal ocean traffic.
Impact on marine life	★★★★★	Offers a natural coastal environment.
Impact on ocean view	★★★★☆	Can block some ocean views, as it can grow up to 30 feet tall in the southernmost parts of Japan and 80 feet in more tropical areas of the world.
Cost	★★★★☆	Considerably less expensive than concrete structures.
Time to build	★☆☆☆☆	Can only grow in the southernmost parts of Japan. They are tropical plants and will not grow in northern parts of Japan. Can take 10-15 years to fully mature.
Maintenance	★★★★☆	Once established, the forest requires limited maintenance to keep it thriving and free of pollution.

Performance of a Pine Forest

H. PINE FOREST		
Criteria & Constraints	**Score**	**Notes**
Ability to break or block waves	★★☆☆☆	Can break up the energy of a small tsunami, but is less effective for larger tsunamis.
Impact on boats/ocean traffic	★★★★★	Does not impact normal ocean traffic.
Impact on marine life	★★★★★	Offers a natural coastal environment.
Impact on ocean view	★★★☆☆	Grows an average of 20 feet tall when planted on a beach, and its branches spread out, making it difficult to see the ocean.
Cost	★★★★★	Inexpensive to build compared to concrete and rock structures.
Time to build	★☆☆☆☆	Can take decades to fully mature.
Maintenance	★★★★☆	Once established, it requires limited maintenance to keep it thriving and free of pollution.

Tsunami Communication Examples

Example 1: Educational flyer from the US National Atmospheric and Oceanic Administration (NOAA)

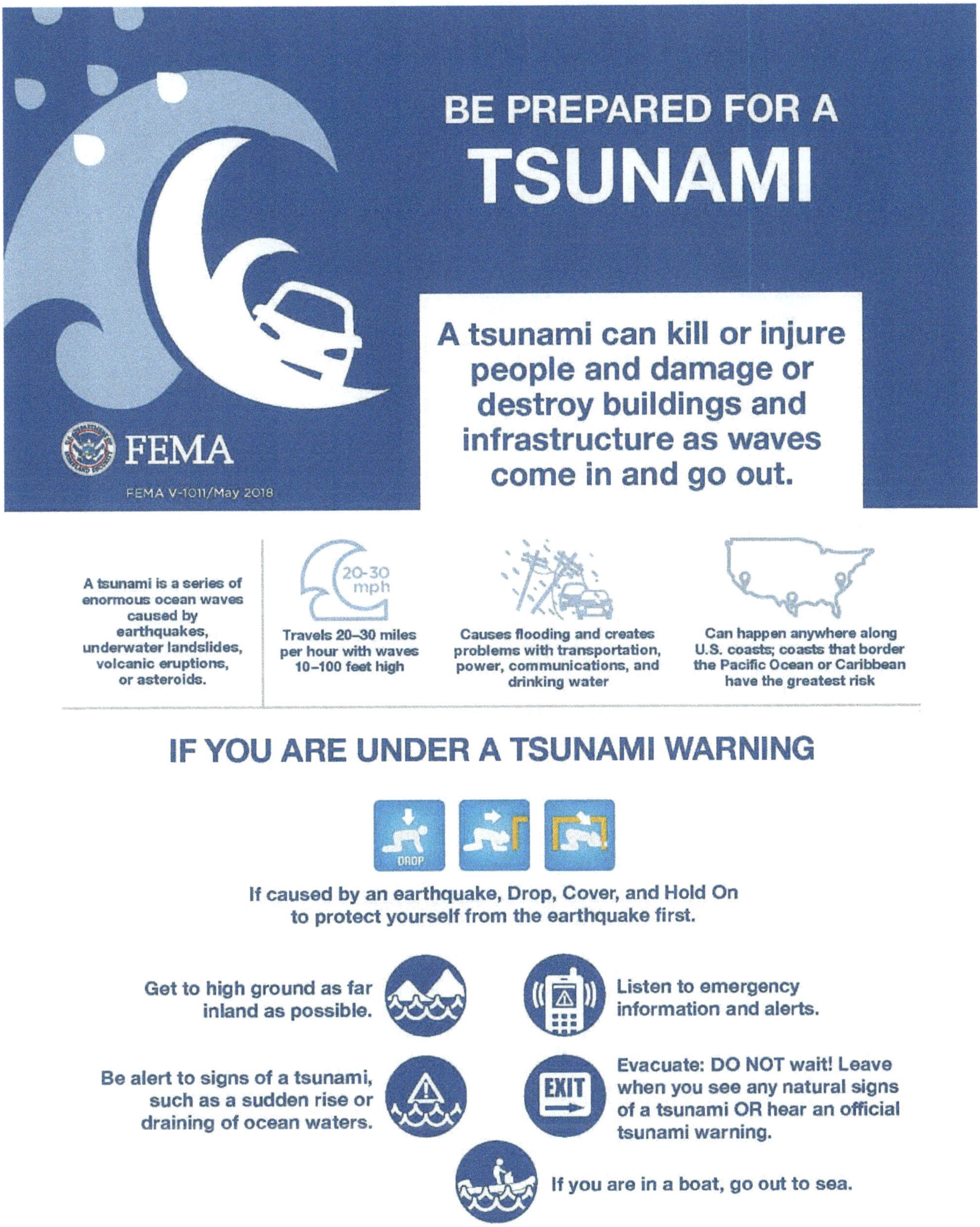

HOW TO STAY SAFE
WHEN A TSUNAMI THREATENS

Prepare NOW

If you live near or visit a coastal area, learn about the tsunami risk. Some at-risk communities have maps with evacuation zones and routes. If you are a visitor, ask about community emergency plans.

Learn the signs of a potential tsunami, such as an earthquake, a loud roar from the ocean, or unusual ocean behavior, such as a sudden rise or wall of water or sudden draining showing the ocean floor.

Know and practice community evacuation plans and map out your routes from home, work, and play. Pick shelters 100 feet or more above sea level or at least one mile inland.

Create a family emergency communication plan that has an out-of-state contact. Plan where to meet if you get separated.

Sign up for your community's warning system. The Emergency Alert System (EAS) and National Oceanic and Atmospheric Administration (NOAA) Weather Radio also provide emergency alerts.

Consider earthquake insurance and a flood insurance policy through the National Flood Insurance Program (NFIP). Standard homeowner's insurance does not cover flood or earthquake damage.

Survive DURING

If you are in a tsunami area and there is an earthquake, first protect yourself from the earthquake. Drop, Cover, and Hold On. Drop to your hands and knees. Cover your head and neck with your arms. Hold on to any sturdy furniture until the shaking stops. Crawl only if you can reach better cover, but do not go through an area with more debris.

When the shaking stops, if there is a warning, either natural signs or an official warning, move immediately to a safe place as high and as far inland as possible. Listen to the authorities, but do not wait for tsunami warnings and evacuation orders.

If you are outside of the tsunami hazard zone and receive a warning, stay where you are unless officials tell you otherwise.

Leave immediately if you are told to do so. Evacuation routes are often marked by a wave with an arrow in the direction of higher ground.

If you are in the water, grab onto something that floats, such as a raft, tree trunk, or door.

If you are in a boat, face the direction of the waves and head out to sea. If you are in a harbor, go inland.

Be Safe AFTER

Listen to local alerts and authorities for information on areas to avoid and shelter locations.

Avoid wading in floodwater, which can contain dangerous debris. Water may be deeper than it appears.

Be aware of the risk of electrocution. Underground or downed power lines can electrically charge water. Do not touch electrical equipment if it is wet or if you are standing in water.

Stay away from damaged buildings, roads, and bridges.

Document property damage with photographs. Conduct an inventory and contact your insurance company for assistance.

Save phone calls for emergencies. Phone systems are often down or busy after a disaster. Use text messages or social media to communicate with family and friends.

Take an Active Role in Your Safety

Go to **Ready.gov** and search for **tsunami**. Download the **FEMA app** to get more information about preparing for a **tsunami**.

FEMA

FEMA V-1011
Catalog No. 17233-12

Example 2: Tsunami flyer in multiple languages

Southern California Earthquake Center/ tsunamizone.org

Southern California Earthquake Center/tsunamizone.org

Example 3: Graphic flyer from Indonesia

Example 4: Cell phone app #1

Japan Tourism Agency website

Japan Tourism Agency website

Example 5: Cell phone app #2

Yuta Takahashi Design Studio

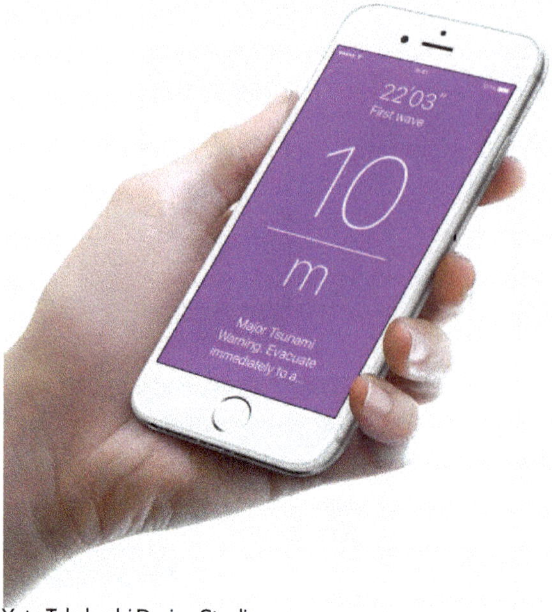

Yuta Takahashi Design Studio

Yuta Takahashi Design Studio

Example 6: Cell phone app #3

sms-tsunami-warning.com

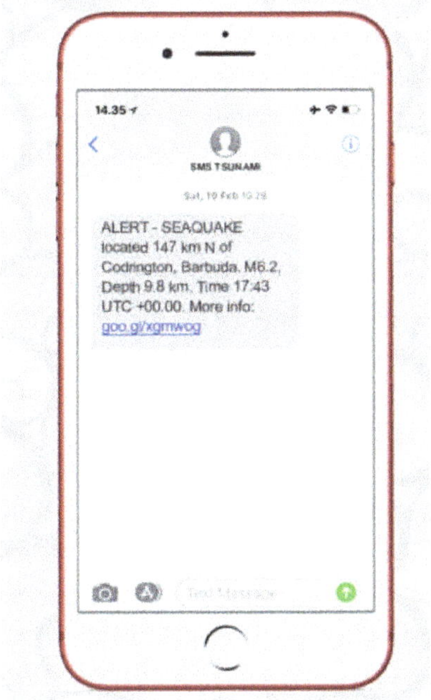

sms-tsunami-warning.com

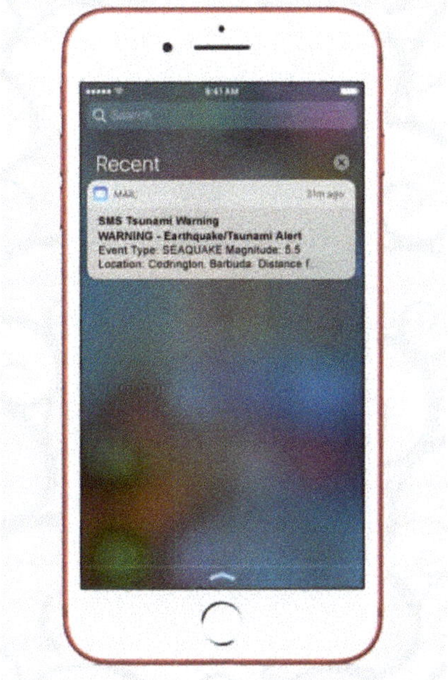

sms-tsunami-warning.com

Hazard Risk Maps

Wildfires

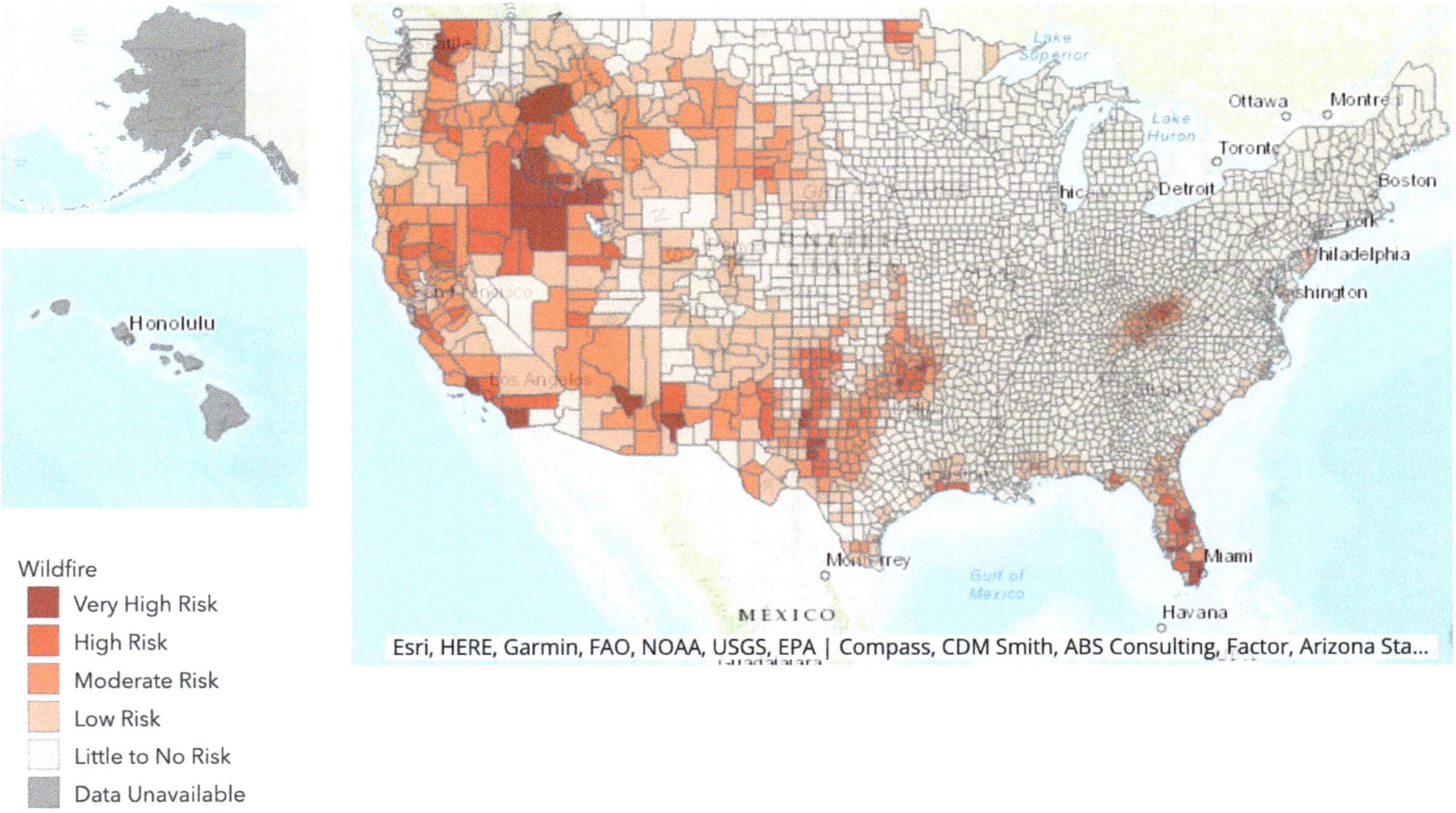

Thunderstorms

Note: Lightning strikes are used as an indicator of thunderstorm risk.

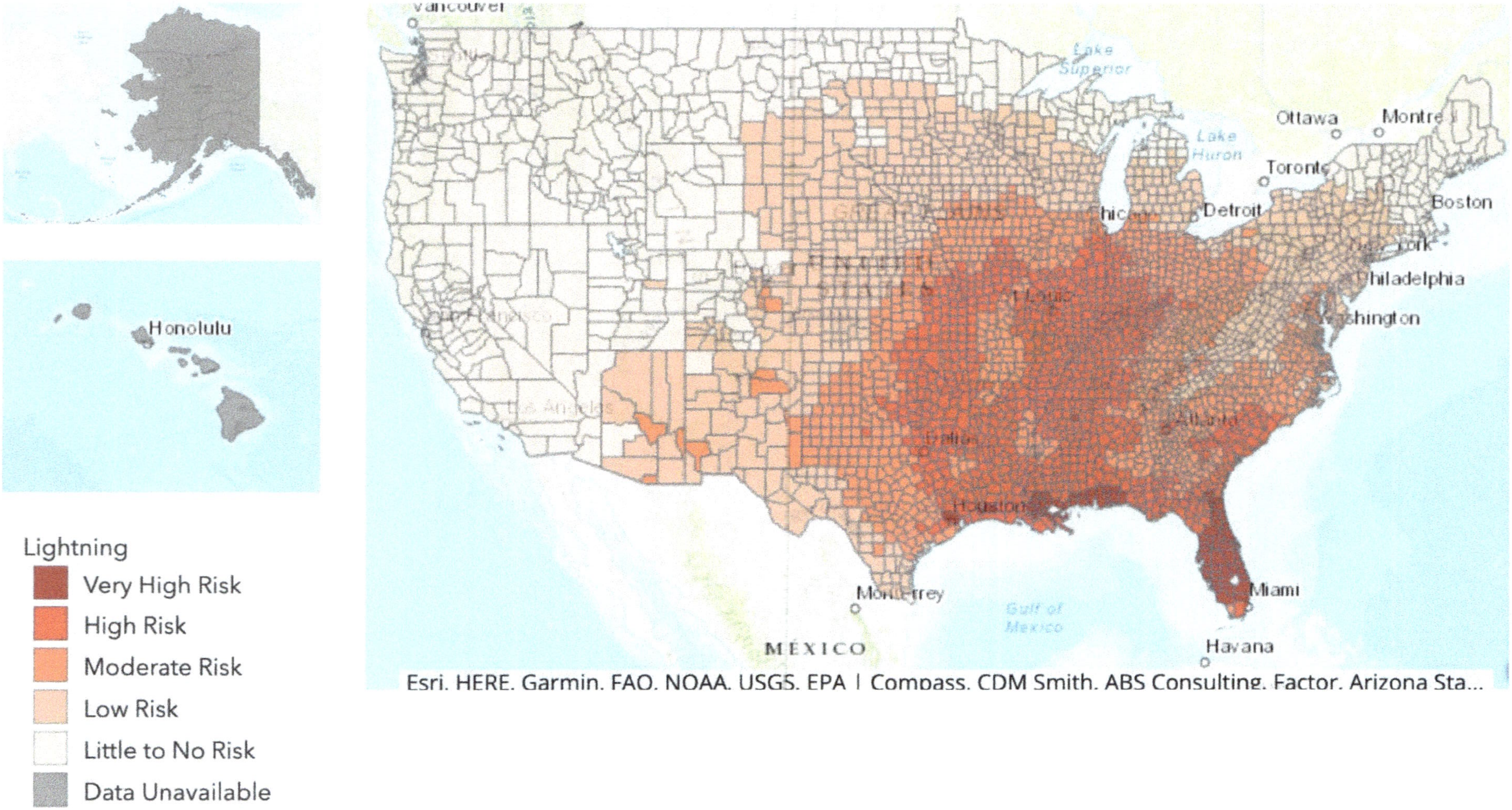

Winter Weather

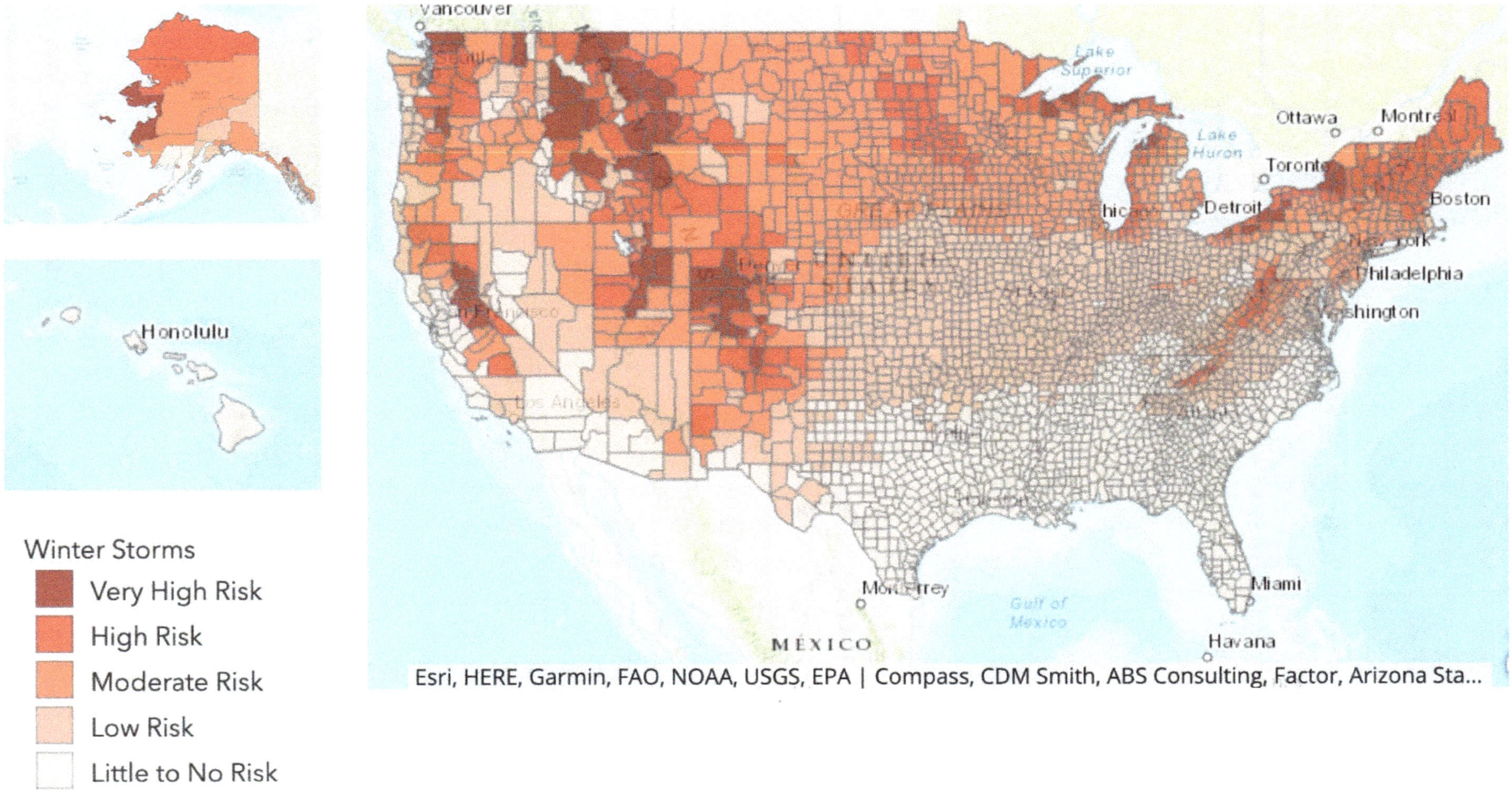

Hurricanes

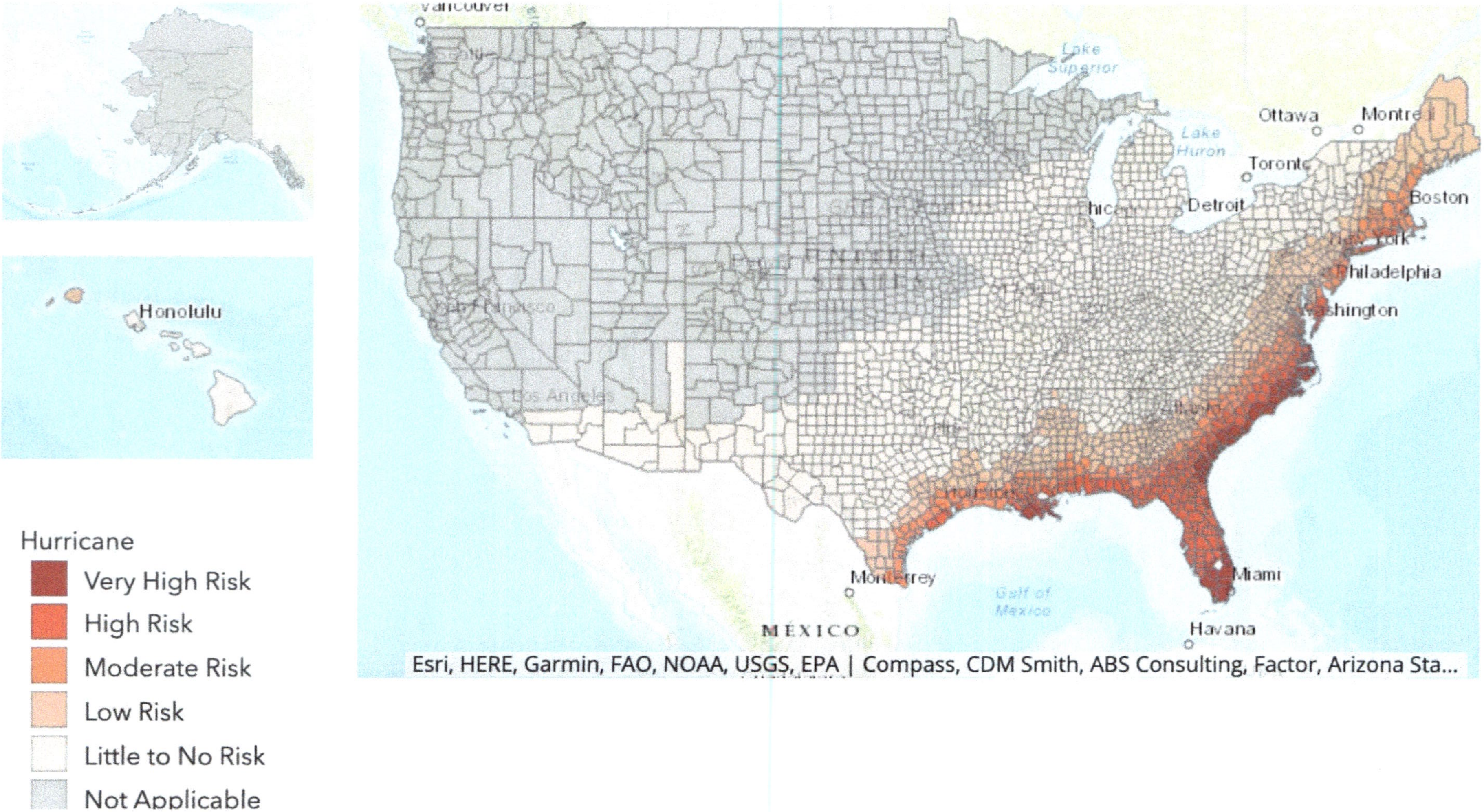
Vancouver
Lake Superior
Lake Huron
Ottawa
Montreal
Toronto
Detroit
Boston
Philadelphia
Miami
Monterrey
Gulf of Mexico
MEXICO
Havana
Honolulu
Esri, HERE, Garmin, FAO, NOAA, USGS, EPA | Compass, CDM Smith, ABS Consulting, Factor, Arizona Sta...
Hurricane
Very High Risk
High Risk
Moderate Risk
Low Risk
Little to No Risk
Not Applicable

Extreme Heat

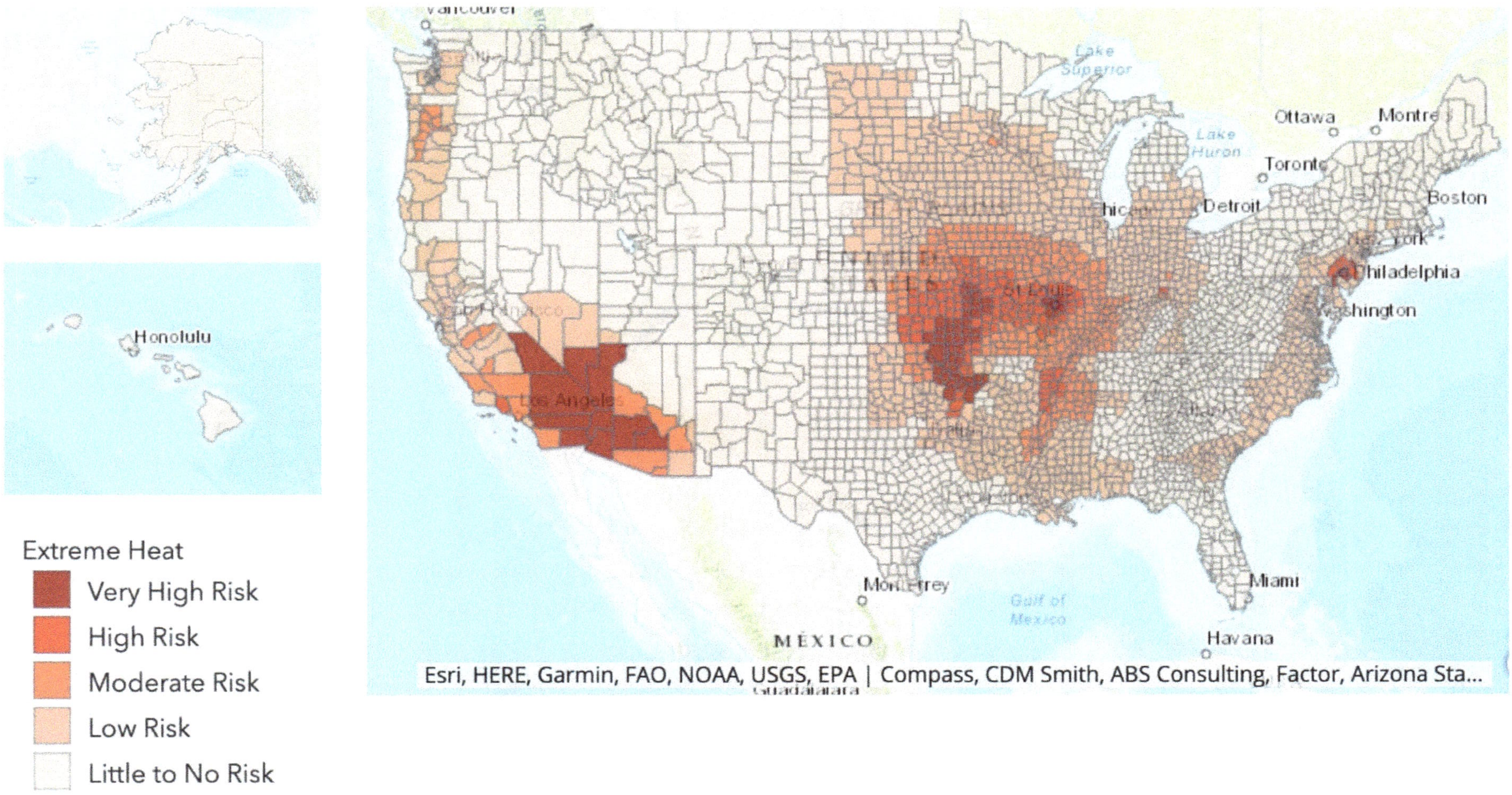

Drought

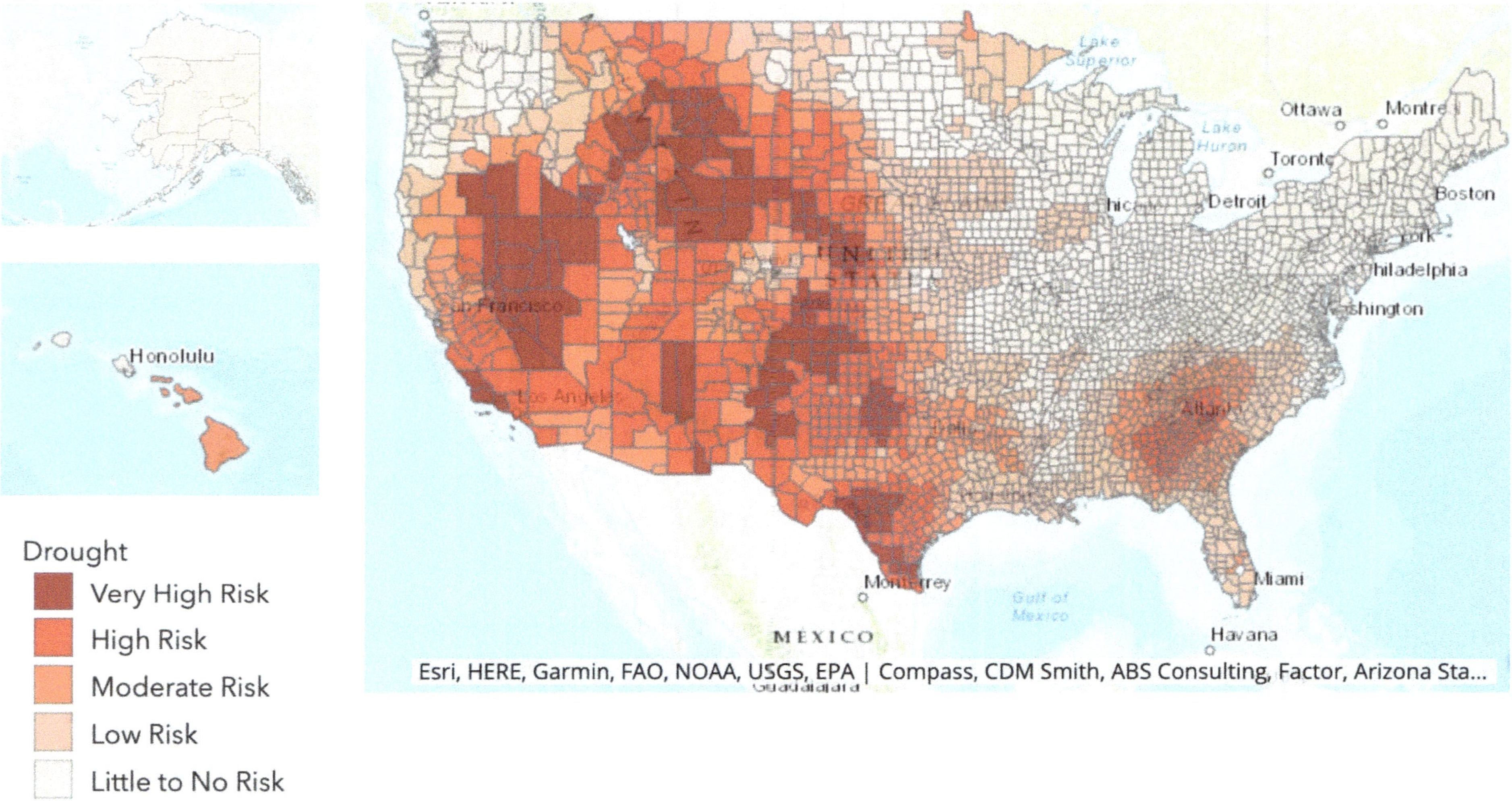
Honolulu
Lake Superior
Lake Huron
Ottawa
Montre
Toronto
Detroit
Boston
Philadelphia
Washington
San Francisco
Los Angeles
Atlanta
Miami
Monterrey
Gulf of Mexico
MEXICO
Havana
Esri, HERE, Garmin, FAO, NOAA, USGS, EPA | Compass, CDM Smith, ABS Consulting, Factor, Arizona Sta...
Drought
Very High Risk
High Risk
Moderate Risk
Low Risk
Little to No Risk

Coastal Flooding

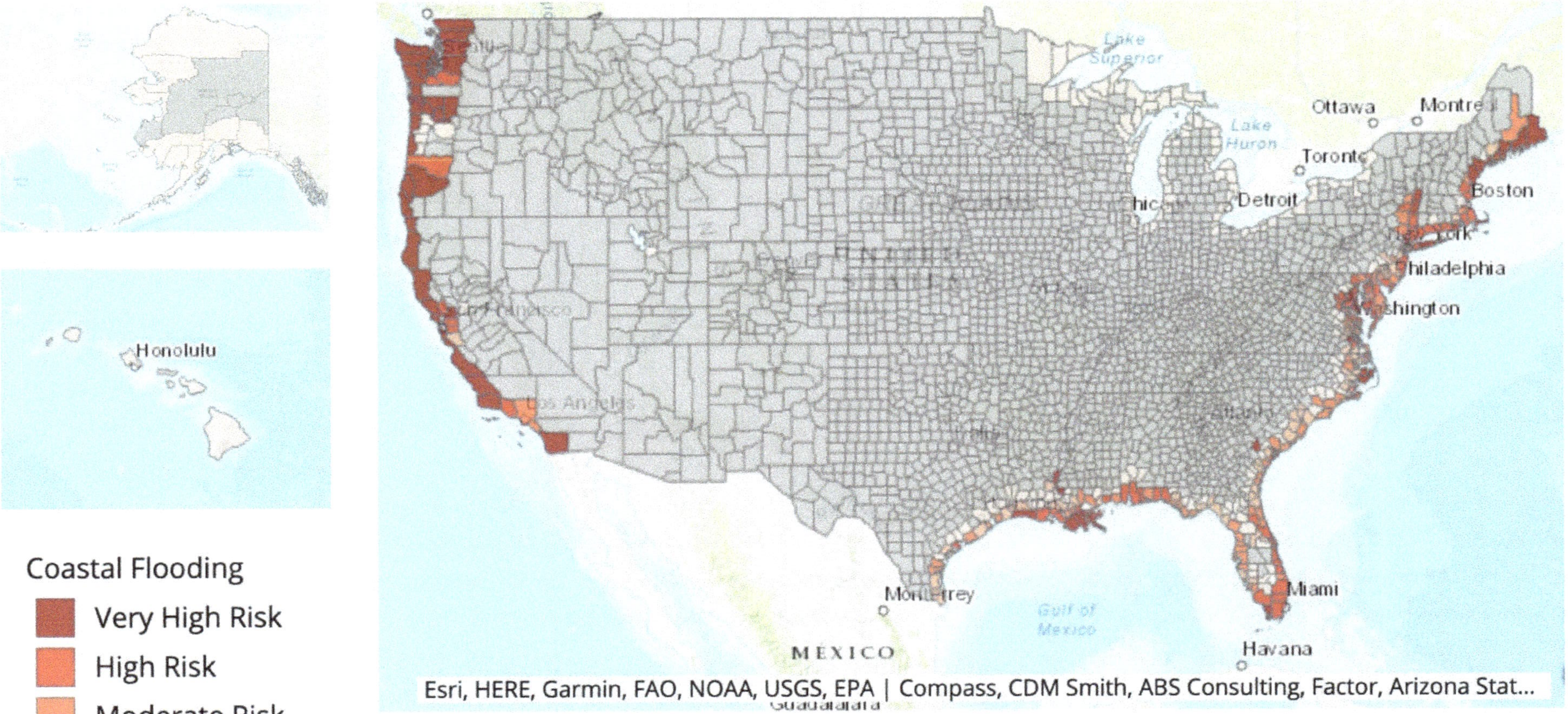

Coastal Flooding

- Very High Risk
- High Risk
- Moderate Risk
- Low Risk
- Very Low to No Risk

Flooding Along Rivers

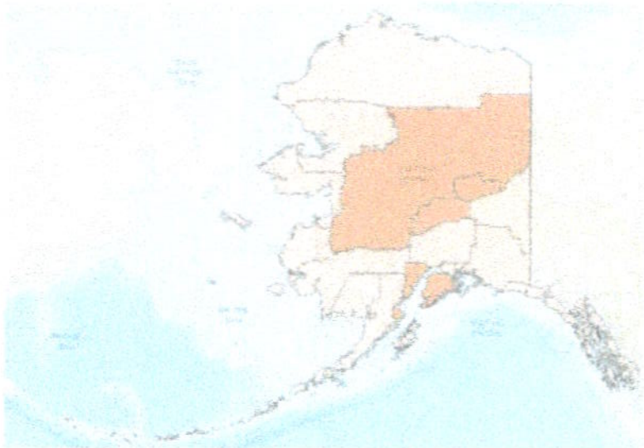

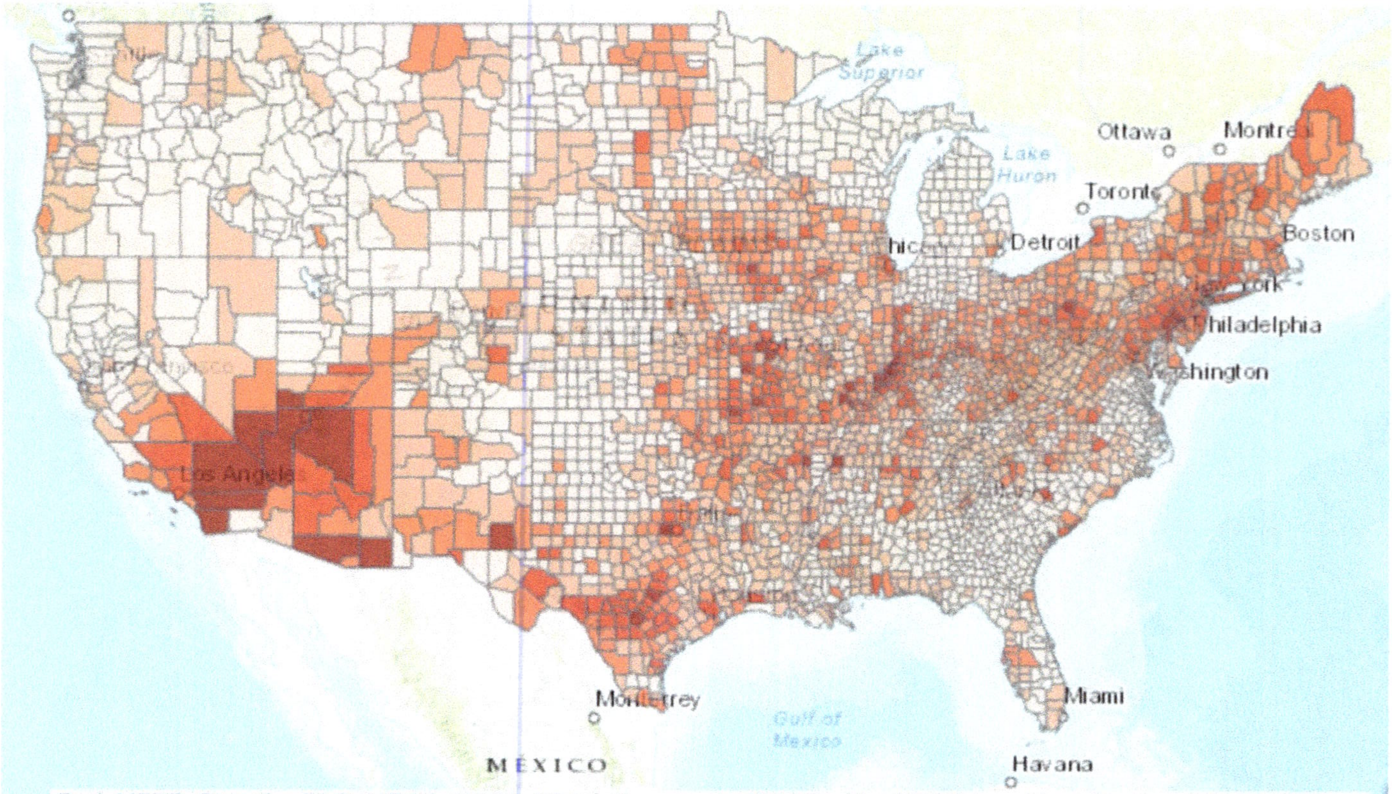

Esri, HERE, Garmin, FAO, NOAA, USGS, EPA | Compass, CDM Smith, ABS Consulting, Factor, Arizona Stat...

Flooding Along Rivers

- Very High Risk
- High Risk
- Moderate Risk
- Low Risk
- Little to No Risk

Earthquake

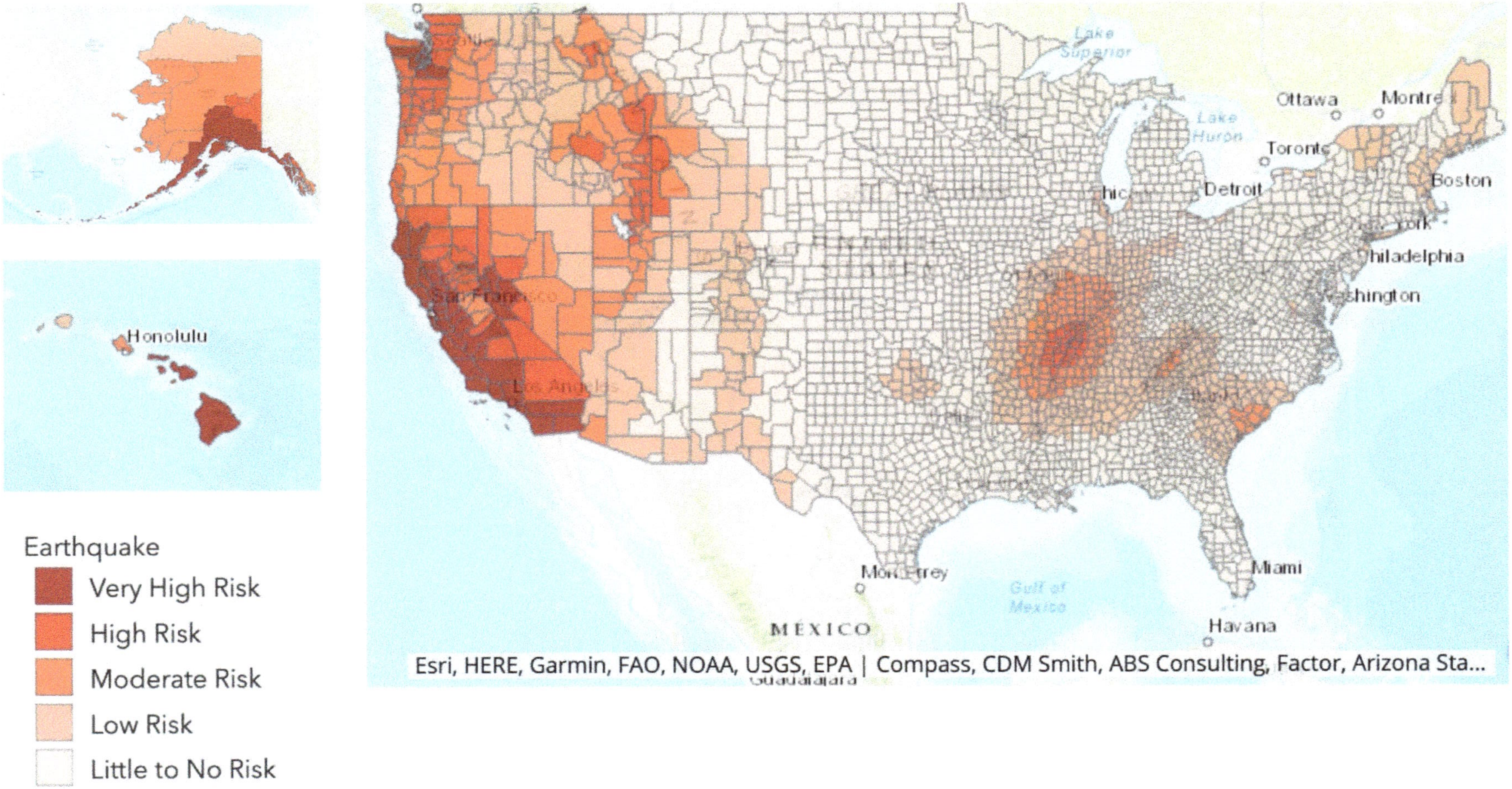

Honolulu
Lake Superior
Lake Huron
Ottawa
Montre
Toronto
Detroit
Boston
Philadelphia
Washington
San Francisco
Los Angeles
Miami
Gulf of Mexico
Havana
MEXICO
Earthquake
Very High Risk
High Risk
Moderate Risk
Low Risk
Little to No Risk
Esri, HERE, Garmin, FAO, NOAA, USGS, EPA | Compass, CDM Smith, ABS Consulting, Factor, Arizona Sta...

Natural Hazards Around the World

About Wildfires

When an uncontrolled fire burns vegetation (trees, brush, or other plants), it is called a wildfire. Sometimes wildfires start if people use fire without care or safety. Other times they start naturally, such as when lightning strikes. Wildfires happen most frequently when a lot of dry vegetation and extreme heat come together. They can spread fast when hot, dry wind is blowing in the area of a fire.

In general, scientists know that wildfires tend to be larger during drier seasons, or droughts. Scientists predict that the amount of land burned by wildfires will increase in the future due to climate change. As global temperatures rise and dry areas become drier, wildfires will burn longer and hotter and destroy more land than before. For the past 20 years in the United States, approximately 5–10 million acres of land have burned due to wildfires each year. In early 2020, Australia experienced very large wildfires that burned 46 million acres of land.

Wildfires can travel up to 14 miles per hour, which is faster than most people can run. They are also very hot. Wildfires can reach temperatures of 1000°F (590°C), which is hot enough to completely destroy nearly everything in its path.

We often think of fire as destructive, but naturally occurring wildfires help many ecosystems. Some trees need fire to release their seeds, and fire can increase the health of a forest by clearing away dry and dead vegetation. Indigenous people in North, Central, and South America have carefully used fire to restore nutrients to soil for centuries.

Helpful Resources

NASA Fire and Smoke Monitoring: https://www.nasa.gov/mission_pages/fires/main/index.html

USDA Forest Service, Active Fire Maps: https://fsapps.nwcg.gov/afm/

NOAA Storm Prediction Center, Fire Weather Outlook: https://www.spc.noaa.gov/products/fire_wx/

FEMA Wildfire brochure: https://www.ready.gov/sites/default/files/2020-03/wildfire-information-sheet.pdf

CDC, Wildfire: https://www.cdc.gov/disasters/wildfires/index.html

Wildfire outside a neighborhood. Michael Held

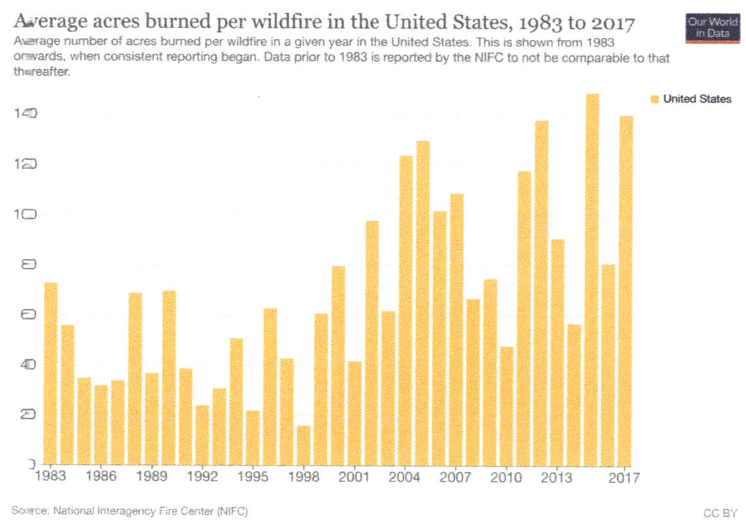

Acres of land burned by year in US wildfires.

Firefighters fighting a wildfire. ELG21

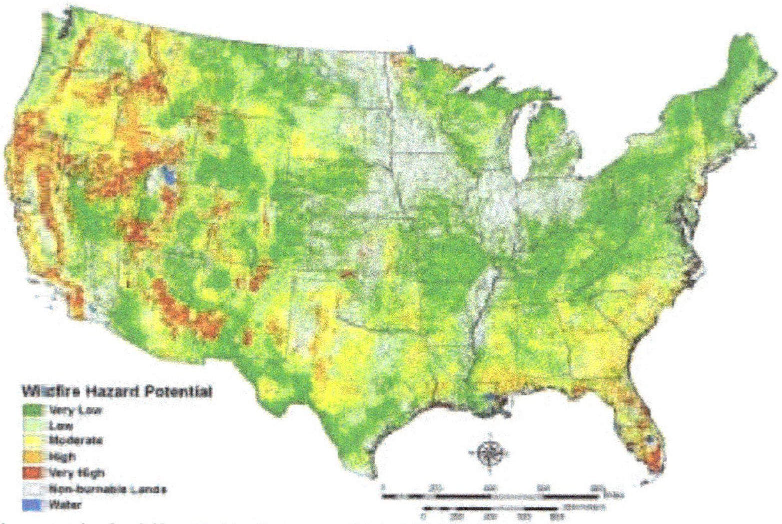

Place at risk of wildfire. Red indicates very high risk USDA

What technologies and design solutions exist to detect, warn, or reduce damage? Wildfire

Forecast and Detect

- Most wildfires are detected by visual contact, such as seeing smoke. They can be observed by people nearby or by aerial surveillance and satellite imagery.
- Wildfires begin suddenly, and when conditions are dry and windy, they spread quickly.
- Scientists can forecast conditions that are favorable for wildfire: warm temperatures, low humidity in the air, high winds, and a source of "fuel" such as trees and brush.
- When these conditions exist, local agencies will put up notifications that the area has "high fire danger." Often, this means no outdoor fires.

Warn and Communicate

- When a wildfire starts, local community firefighters and law enforcement often coordinate warnings and evacuations with neighborhoods most at risk.
- If you live in a place at risk of wildfire, have an evacuation bag packed so that your family can evacuate quickly if you must do so.
- The Emergency Alert System (EAS) and National Oceanic & Atmospheric Administration (NOAA) Weather Radio provide wireless alerts to people's cell phones.
- Often communication includes warnings about air quality and how to protect yourself when going outside. Smoke from wildfires can spread out over large areas, affecting more residents than the fire alone.

Reduce Damage

- Use fire-resistant materials to build or repair homes.
- Create a fire-resistant zone free of leaves, debris, or flammable materials for at least 30 feet away from home.
- Shut off all gas and propane to the house.
- Wet your roof and areas at least 50 feet away from home to keep air moisture high.
- Have a plan for pets and any livestock or animals.

About Severe Thunderstorms

A short-lived rain shower that produces lightning is called a thunderstorm. In the United States, thunderstorms usually form when dry, cold air from the north meets with warm, moist air from the south. These conditions commonly occur in late spring and early summer. Some thunderstorms are severe and can produce winds of over 50 miles per hour, hail, tornadoes, and lightning. Hailstones are chunks of ice that fall from strong thunderstorms. A tornado is a column of rotating air that is in contact with the ground. Even though they are made of air, they are visible because you see the dust and debris that are caught in them.

Conditions associated with thunderstorms:

- *Tornadoes* occur most often in the Great Plains and Midwest, but they also occur in other parts of the country. Tornadoes move very fast (up to 50 miles per hour) and can destroy or cause severe damage to structures on the ground.
- *Hail* does about 1 billion dollars in damage to crops and property each year in the United States. The National Weather Service classifies hail as "severe" if the hailstones have a diameter of 1 inch or more, but smaller hail can damage crops by bruising fruit meant to be sold to a store or restaurant.
- Humans are involved in *lightning strikes* over 200,000 times a year. Most lightning victims survive but often have long-term health issues, such as hearing damage. About 44 people die from lightning strikes each year in the United States. Because Florida experiences more thunderstorms than any other state, it has the most lightning fatalities each year.

Depending on the type of storm, different notifications can be issued at different times. Here are what some of the notifications mean:

- **Severe Thunderstorm Watch:** A severe thunderstorm watch indicates that conditions are favorable for the formation of severe thunderstorms.
- **Severe Thunderstorm Warning:** A severe thunderstorm warning means that the area has or will have severe thunderstorm conditions. This includes wind speeds of 58 miles per hour or greater and/or hail that is 1 inch in diameter or larger.
- **Tornado Watch:** A tornado watch is issued when weather conditions could lead to tornadoes. A watch can cover parts of a state or several states.
- **Tornado Warning:** A tornado warning is issued when a tornado has been reported by spotters or has shown up on radar images. A warning will be issued for the area that would be in the path of the tornado and means people and animals should find a safe place.

Helpful Resources

NOAA Severe Storms Laboratory :https://www.nssl.noaa.gov/education/svrwx101/
North Carolina Climate Office, Severe Weather Hazards: https://climate.ncsu.edu/edu/SevereHazards
FEMA Storms and Lightning: https://www.fema.gov/media-library-data/2607c3fe71a68fe165a53ec189fba37e/FEMA_FS_thunderstorm_508.pdf
FEMA Tornadoes: https://www.fema.gov/media-library-data/20130726-1622-20490-7783/tornadoesfactsheet_final.pdf

Thunderstorm with lightning strike. NOAA

Thunderstorm with tornado. Nikolas Noonan

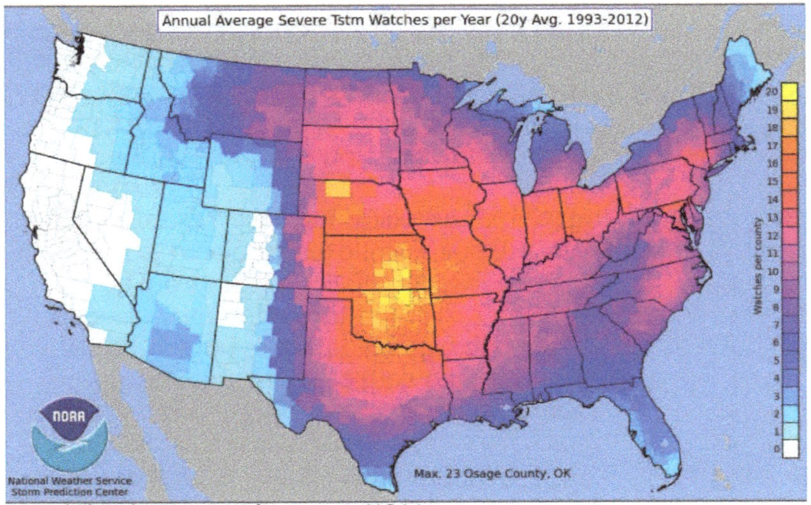

Annual Thunderstorm Watches per year. NOAA

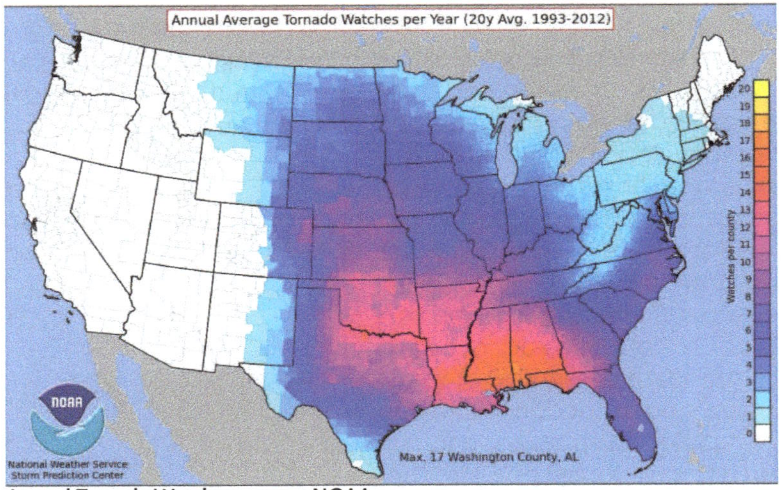

Annual Tornado Watches per year. NOAA

What technologies and design solutions exist to detect, warn, or reduce damage? Thunderstorms

Forecast and Detect

- Thunderstorms are common worldwide and at all times of year. There are about 100,000 thunderstorms in the United States each year alone, but only 10% of these become severe.
- Satellites track clouds and their data and send that information to meteorologists and the National Weather Service.
- Data on the storm's wind speed, direction, amount of precipitation, density, and movement help meteorologists determine what might be happening.
- Lightning, hail, strong winds, and circular cloud rotation data is sensed by radar and satellites. Computers and storm experts analyze this data and issue warnings.
- Watch for tornado danger signs: dark, often greenish clouds—a phenomenon caused by hail; wall cloud—an isolated lowering of the base of a thunderstorm; cloud of debris; rotating funnel-shaped clouds; loud roar.

Warn and Communicate

Get a weather radio or app and know your alert sounds and sirens:

- Weather radios
- Phone apps
- TV
- Radio
- Tornado sirens
- Local emergency management (police, fire, and other government organizations) patrol areas with sirens on to help alert families, if necessary.
- Take cover—find the sturdiest, closest shelter and go inside during a severe storm. Protect yourself with blankets or your arms.
- Avoid flooded roads.
- Do not touch metal.

Reduce Damage

- Bring outdoor items like patio furniture and plants inside, so they do not become flying objects.
- Add extra support to roofs and check to make sure sheds are bolted to the concrete.
- Use hail-resistant siding, storm windows, and roofing materials.
- Cut down or trim trees that may be in danger of falling on your home.
- Consider buying surge protectors or a lightning protection system to protect your home, appliances, and electronic devices.
- Unplug electrical appliances.

About Winter Storms

Winter storms are combinations of very cold temperatures, large amounts of frozen precipitation, and high winds. There are several types of winter storms:

- **Blizzards** are dangerous winter storms that come with blowing snow and wind, making it hard to see. They often have heavy snowfalls and severely cold temperatures.
- An **ice storm** is a storm that places at least .25" of ice on surfaces. This can be hazardous for driving and walking. Tree branches and powerlines can easily snap under the weight of the ice.
- **Lake effect storms** are cold, dry air masses that move over the Great Lakes regions and pick up a lot of moisture from the water. The water in the air becomes snow, and the storm dumps the snow in areas generally to the south and east of the lakes.
- **Snow squalls** are brief, intense snow showers that come with strong, gusty winds.

Cold air, moisture, and lift are all needed to form a winter storm. Lift happens when cold and warm fronts meet, or as air travels up a mountainside. These storms happen every year in the United States from mid to late fall through late spring.

Winter storms and blizzards can last anywhere from a few hours to several days. They can be small and concentrated in one area, or can be large and stretch across several states. These storms can knock out heat, power, and communication to homes and businesses. People who leave home during winter storms are at risk for car accidents, hypothermia, and frostbite. People who heat their homes using gas are at a higher risk of carbon monoxide poisoning.

Depending on the conditions, different notifications can be issued at different times. Here are what some of the notifications mean:

- **Winter Weather Advisory:** Winter weather is expected, but it is not going to be hazardous enough to meet warning criteria.
- **Winter Storm Watch:** A significant winter storm could happen. Watch out for weather conditions to change.
- **Winter Storm Warning:** Life-threatening, severe winter conditions have begun or will begin within 24 hours.
- **Blizzard Warning:** Winds and wind gusts of 35 miles per hour or more are happening right now. There is a lot of falling or blowing snow that makes it impossible to see more than a quarter of a mile. This is expected to happen for 3 hours or longer.

Helpful Resources

NOAA Severe Storms Laboratory: https://www.nssl.noaa.gov/education/svrwx101/
Ready.gov, Winter Storms: https://www.ready.gov/winter-weather
FEMA: https://www.fema.gov/media-library-data/2607c3fe71a68fe165a53ec189fba37e/FEMA_FS_thunderstorm_508.pdf
CDC, Winter Storms: https://www.cdc.gov/disasters/winter/index.html

Snow plow clearing snow in blizzard. Eric Mlean

Vehicles covered in snow. Patino Jhon

Winter storm in a neighborhood. Eric Mlean

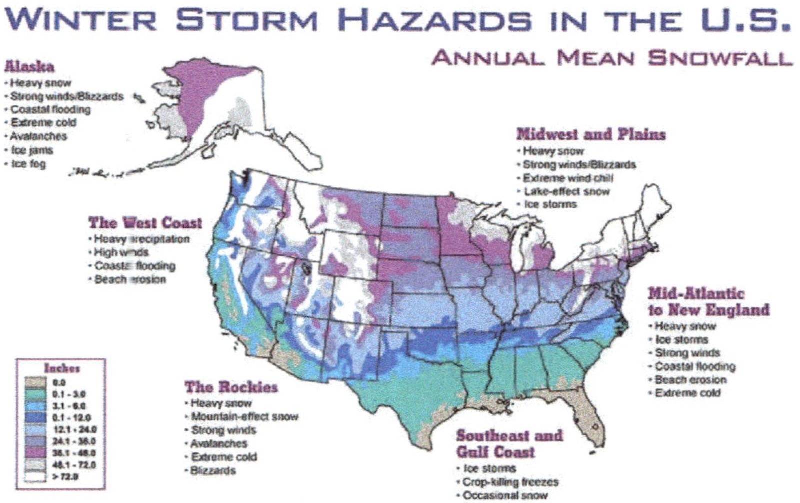

Map of Winter Storm Chances in US. National Weather Service

What technologies and design solutions exist to detect, warn, or reduce damage? Winter Storms

Forecast and Detect

The United States has satellite and radar imaging, along with hundreds of weather stations throughout the United States. The sensors track air conditions and alert meteorologists when conditions are right for a winter storm to occur. Generally, people have several days of warning based on this detection system.

Warn and Communicate

If your National Weather Service local location detects a winter storm, they will send out a watch or a warning for your area. Some ways people are notified include the following:

- Cell phone alerts and apps
- Radio alerts
- TV alerts
- Newspapers (if known far enough in advance)

Reduce Damage

- Many cities plow or clear roads and have response systems in place to keep roads clear of snow to reduce hazardous driving conditions.
- Homes should use good insulation to protect water pipes (from bursting) and keep the home warmer during winter storms.
- Cut tree branches away from power lines.

About Hurricanes

Rotating storm systems that develop over warm ocean water are called tropical depressions or tropical storms. Once the winds inside reach 74 miles per hour or more, it is called a hurricane. This type of storm has different names, depending on where it forms: hurricanes form in the North Atlantic and Northeast Pacific Oceans; cyclones form over the South Pacific and Indian Oceans; and typhoons form over the Northwest Pacific Ocean.

Some hurricanes are small and others are large. The smallest hurricane recorded was only 22 miles across. The largest (Typhoon Tip) was over 1,250 miles wide. The strength of a hurricane can change as time goes on. Scientists use wind speed to classify the strength of a hurricane on the Saffir-Simpson scale:

- **Category 1:** Winds range from 74–95 miles per hour, making them as fast as a Major League Baseball throw. Can cause damage to roofs, shingles, gutters, and siding. Large branches of trees and some full trees may fall. Power lines will be damaged, leaving people without electricity for several days. The storm surge is generally 4–5 feet above normal.
- **Category 2:** Winds range from 96–110 miles per hour, making them as fast as a fastball pitch. Causes major roof and siding damage, trees are snapped or uprooted, and power outages can last days to weeks. The storm surge is generally 6–8 feet above normal.
- **Category 3:** Winds range from 111–129 miles per hour, making winds as fast as a professional tennis serve. Homes suffer major damage, and roofs will be removed. Trees will be uprooted or snapped, electricity and water will be unavailable for days or weeks after. Storm surge is generally 9–13 feet above normal. Some roads may be damaged or blocked by trees.
- **Category 4:** Winds range from 130–156 miles per hour, making them faster than the world's fastest roller coaster. Homes can lose roofs and some outside walls. Most trees will be uprooted and power poles will come down. Power outages and water will last weeks or months, making the areas unsafe until the repairs are made. Storm surge is generally 13–18 feet above normal. Most Category 4 storms come with orders to evacuate the coastal region as far as 6 miles inland. Some roads and bridges are damaged.
- **Category 5:** Winds range at or above 157 miles per hour, making them as fast or faster than a skydiver falling head first, or a high-speed train. Most homes and trees are completely destroyed. Power and water will be out for weeks or months, making the area unsafe to live in. All Category 5 storms come with evacuation orders. Storm surge will be 19 or more feet above normal. Evacuations are generally required for those living 5–10 miles from the shoreline. Damage to roads, bridges, and other structures cause safety issues for the communities.

Depending on the conditions, different notifications can be issued at different times. Here are what some of the notifications mean:

- Hurricane Watch: hurricane conditions (winds of 74 miles per hour or more) are possible in the area.
- Hurricane Warning: hurricane conditions (winds of 74 miles per hour or more) are expected in the area.

Helpful Resources

Ready.gov, Hurricanes: https://www.ready.gov/hurricanes
FEMA factsheet: https://www.fema.gov/pdf/hazard/hurricane/2010/hurricane_week_preparedness_factsheet_ready.pdf
CDC, Hurricanes: https://www.cdc.gov/disasters/hurricanes/index.html
Video explaining why hurricane predictions are so variable: https://www.youtube.com/watch?v=e1ilcGGsBu8

Hurricane from space. WikiIImages

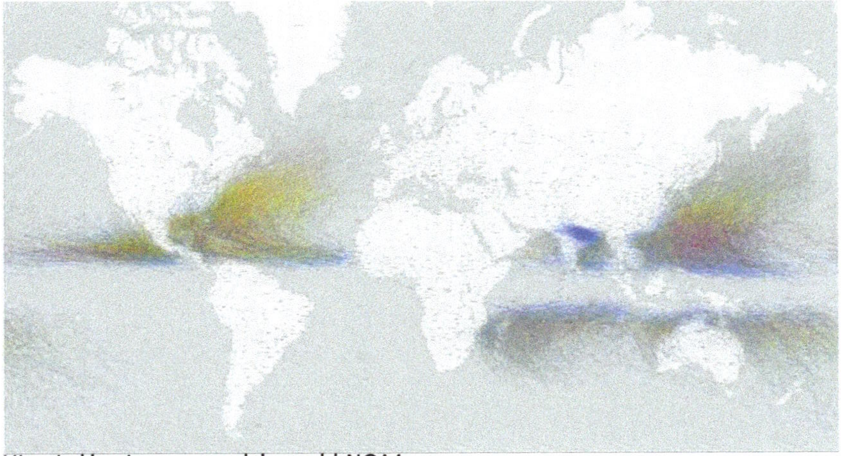
Historical hurricanes around the world. NOAA

Storm surge from a hurricane. Wickedgood

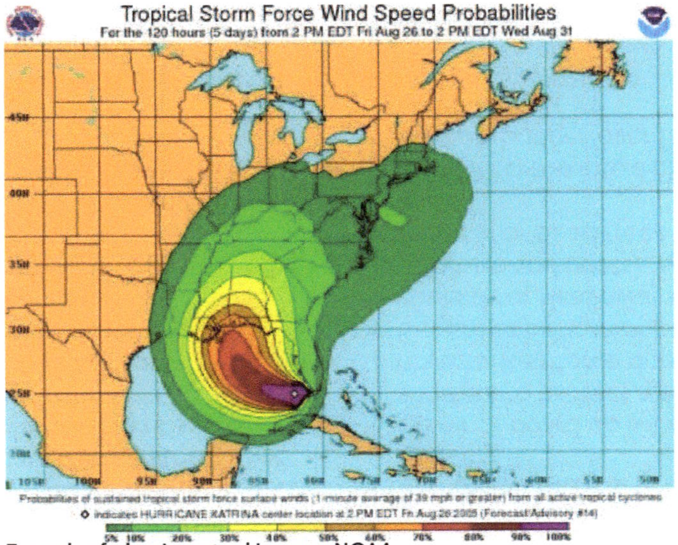

Example of a hurricane tracking map. NOAA

What technologies and design solutions exist to detect, warn, or reduce damage? Hurricanes

Forecast and Detect

- Hurricane season for the United States is from June 1st through September 30th; 97% of hurricanes occur during this time.
- Storms that can form hurricanes can be detected roughly 5 days in advance.
- Satellites collect data about the storm.
- Drifter buoys send data about ocean currents and sea levels to scientists.
- Planes drop special tools into the hurricane to collect additional data.
- Supercomputers use all the data to create models of the hurricane path and severity.

Warn and Communicate

- People normally receive 2 days of warning to prepare for the storm.
- A potential path of the hurricane is predicted by scientists
- Meteorologists evaluate and communicate information to the community and government.
- Watches and warnings are given for the storm.
- Storm surge watches/warnings are issued to help coastal communities prepare for potential flooding.

Reduce Damage

- Sandbags can help prevent flooding.
- Cover or board up windows with shutters to protect against flying objects. Reinforce doors.
- Secure or bring in objects that could fly or float away (i.e., outdoor furniture and boats).
- Trim old or dead trees and tree limbs.
- Do NOT climb into a closed attic.
- Stay off bridges over fast moving water.

About Extreme Heat

Extreme heat is a period of high heat and humidity for at least 2–3 days. These events are caused when high pressure air in the upper atmosphere traps heat at the surface of the Earth.

A certain temperature and humidity level might be considered extreme in one area but typical in another. This means the definition of extreme heat varies by location. For example, 86°F high temperatures in Anchorage, Alaska broke a record in 2019, but the same temperature is common in parts of Florida. Extreme heat events can last for several days without a break in the temperatures. In some places, temperatures could exceed 130°F. Heat waves are becoming more common today than they were in the past.

Extreme heat events can develop quickly in an area, but they are often predicted several days ahead of time. This can give people time to prepare. Depending on the conditions, different notifications can be issued at different times. Here are what some of the notifications mean:

- **Excessive Heat Warning**: issued within 12 hours of the onset of extremely dangerous heat conditions. The general rule of thumb for this warning is when the maximum heat index temperature is expected to be 105° or higher for at least 2 days and nighttime air temperatures will not drop below 75°. However, these criteria vary across the country, especially for areas not used to extreme heat conditions.
- **Excessive Heat Watch**: issued when conditions are favorable for an excessive heat event in the next 24 to 72 hours. A watch is used when the risk of a heat wave has increased but its occurrence and timing is still uncertain.
- **Heat Advisory**: issued within 12 hours of the onset of extremely dangerous heat conditions. The general rule of thumb for this advisory is when the maximum heat index temperature is expected to be 100° or higher for at least 2 days, and nighttime air temperatures will not drop below 75°. These criteria vary across the country, especially for areas that are not used to dangerous heat conditions.
- **Excessive Heat Outlook**: issued when the potential exists for an excessive heat event in the next 3–7 days. An outlook provides information to those who need considerable lead-time to prepare for the event.

Helpful Resources

Ready.gov, Extreme Heat: https://www.ready.gov/heat
EPA, Extreme Heat: https://www.epa.gov/sites/production/files/2016-10/documents/extreme-heat-guidebook.pdf
CDC, Extreme Heat: https://www.cdc.gov/disasters/extremeheat/index.html

Woman cooling down with a fan. timole20

Rooftops painted white in New York City. NASA's Earth Observatory

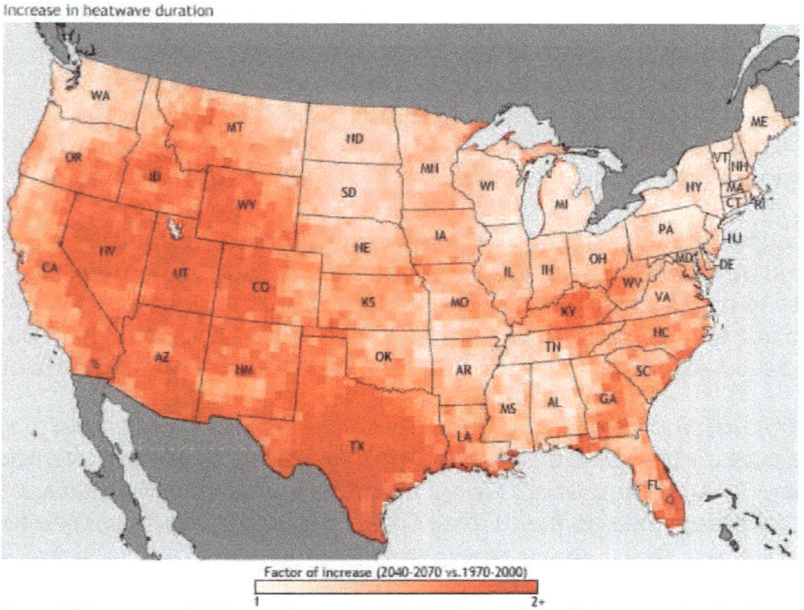

Potential increases in heatwave duration in 2040-2070. NOAA Climate.gov map by Dan Pisut and Richard Rivera, based on data provided by Mary Jo Nath and Gabriel Lau

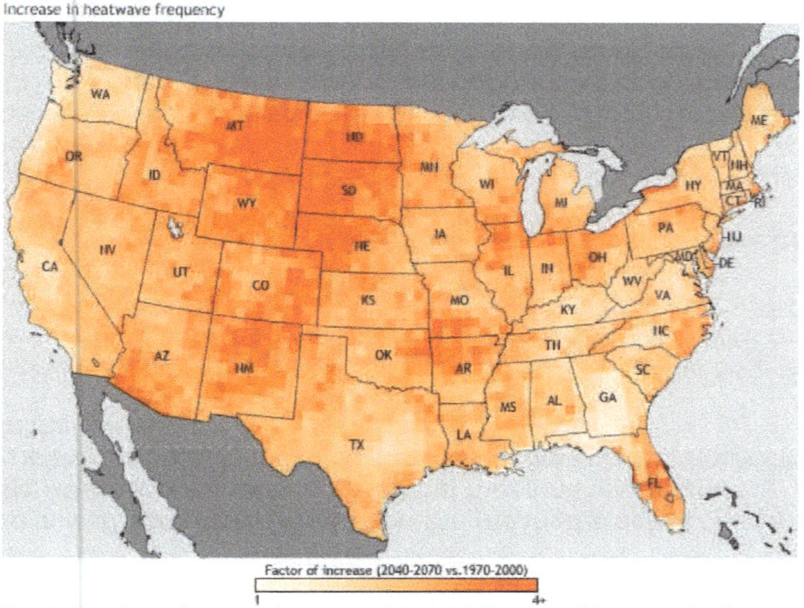

Potential increase in heatwave frequency in 2040-2070. NOAA Climate.gov map by Dan Pisut and Richard Rivera, based on data provided by Mary Jo Nath and Gabriel Lau

What technologies and design solutions exist to detect, warn, or reduce damage? Extreme Heat

Forecast and Detect

- Extreme heat tends to happen in the summer months, but there is no official "heat season."
- Satellites actively monitor and send data about temperature and precipitation changes all over the world.
- Meteorologists can use temperature and weather patterns from all over the world to forecast heat waves.
- Heat waves can be forecast about 10 days from when they will occur.
- If the heat wave comes from over an ocean, sometimes the heat wave can be forecasted as much as 6 weeks in advance.

Warn and Communicate

- Extreme heat advisories, watches, and warnings are broadcast over television and radio.
- Push notifications can also go out on cell phones and other devices.
- Weather apps also keep up-to-date information regarding the temperature for easy access by anyone with a device.

Reduce Damage

Using what we know about light and energy, we know dark surfaces will be warmer than lighter surfaces during an extreme heat event. Some cities are painting dark surfaces white to reduce the amount of energy transferred to the surface. Roads, train tracks, and roofs are some of the many things painted white to help cool cities and reduce temperatures. Green spaces, such as parks and areas with trees, can also provide cooler areas within cities.

About Droughts

Droughts are long periods without water or access to water in an area. Droughts occur because of a lack of precipitation, groundwater, or other water sources in an area. There is no specific drought season. Many times droughts are part of heat waves, which dry out areas and cause a water shortage. See About Extreme Heat for more details.

Droughts can cover areas from the size of a county to an entire country. They can last from weeks to years depending on the size and location of the drought. If farming and ranching areas are without water for long periods of time, it can result in crop damage, loss of livestock, and loss of income for many farmers. This results in reduced access to food for people who live in the area and people who buy food produced there. Droughts can vary in how much they impact an area. If water resources are well maintained, the impact of a drought can be less dramatic. The most widely discussed drought in United States history is the Dust Bowl, which occured in the 1930s and ruined millions of acres of farmland.

The United States Drought Monitor is a map released weekly that shows both the intensity and impact of drought conditions in the United States. Drought intensity is measured on a scale from D0 (abnormally dry) to D4 (exceptional drought). Impact is described as either short-term or long-term.

Helpful Resources

Ready.gov, Drought: https://www.ready.gov/drought
US Drought Monitor: https://droughtmonitor.unl.edu/

Cracked ground from lack of moisture. Jose Antonio Alba

Rain barrel on the side of a building.
Harry Cunningham

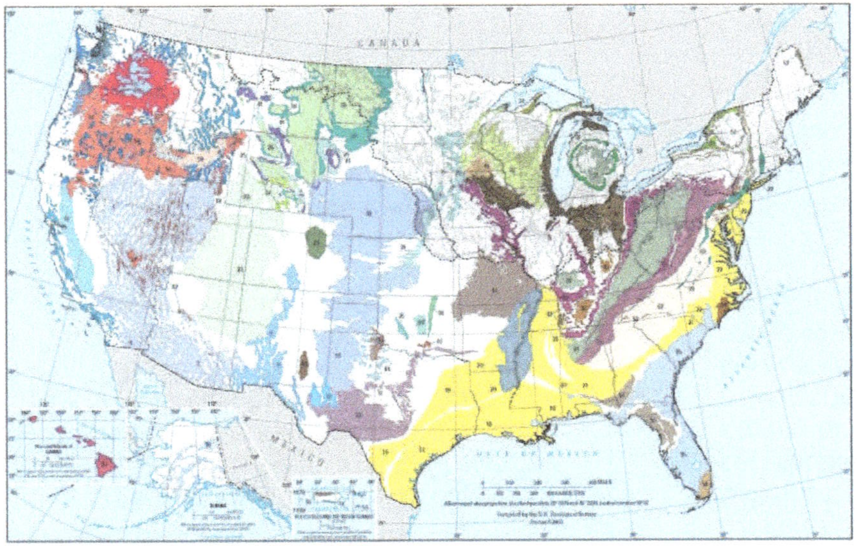

Map of aquifers in the United States. USGS

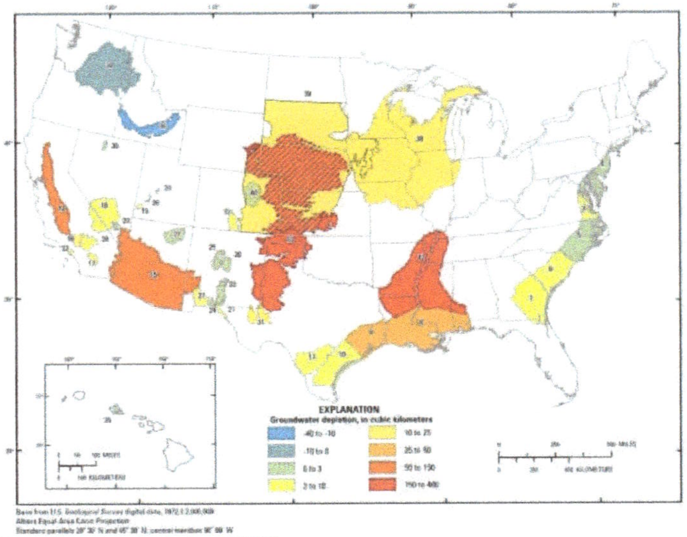

Map of aquifer depletion. USGS

What technologies and design solutions exist to detect, warn, or reduce damage?

Forecast and Detect

- The National Integrated Drought Information System (NIDIS) watches different areas for signs of drought.
- The Drought Early Warning Systems (DEWS) is monitored to determine if a drought might happen.
- Seasonal drought and climate outlooks are published by climate scientists.
- Satellites monitor the amount of evaporation in an area, which can provide early warning signs of fast-occurring droughts.

Warn and Communicate

- Scientists communicate drought conditions and risks to local emergency management and new agencies.
- Local governments and news agencies inform the public.

Reduce Damage

- Aquifer storage: Water from the surface is stored in aquifers during a drought. This water can be used during and after a drought.
- Check plumbing for breaks and leaks to conserve water when drought conditions exist.
- Homes can reduce water usage with conservative appliances, such as water efficient toilets and shower heads.
- Increase awareness and support for water conservation efforts.
- Cover pools and spas to reduce water loss.

About Flooding

When large amounts of water overflow onto land that is normally dry, it is called flooding. Flooding can happen at any time of the year.

Floods happen for several different reasons. Floods that are not near coasts are usually caused by weather events like heavy rain or melting snow, which causes rivers and creeks to overflow. Coastal flooding can be caused by these, too, but also by storm surges from hurricanes and large ocean waves like tsunamis.

Areas at a lower elevation ("low-lying") and near a source of water, like a creek, river, or lake, flood before higher ground. Some floods happen more slowly, and some floods happen more quickly. When water rises very quickly and people have very little time to respond, it is called a "flash flood."

Sometimes floods can be only a few inches of water on the ground, but in some places it can reach as high as the tops of roofs on houses. Flooding often damages homes and businesses, causes power outages, and disrupts transportation networks.

Flooding is the most common natural hazard in the United States, and often the costliest. Because all parts of the country are at risk of flooding, the federal and state governments, and local communities, have developed many solutions to prevent or reduce flood damage and also protect people when it happens.

Depending on the conditions, different notifications can be issued at different times. Here are what some of the notifications mean:

- Flood watch: A flood watch means a flood may occur in an area. Monitor conditions and prepare to respond if a warning occurs.
- Flood warning: A flood warning means a flood will occur in the area and to respond appropriately.

Helpful Resources

Research your own flood risk using the National Flood Maps Service: https://msc.fema.gov/portal/home
CDC, Floods: https://www.cdc.gov/disasters/floods/index.html
Ready.gov, Floods: https://www.ready.gov/floods

A dam holds back floodwaters. Hans Braxmeier

Cars are submerged under flood waters. Chris Gallagher

A road was washed out by floodwaters. By permission of KMTV

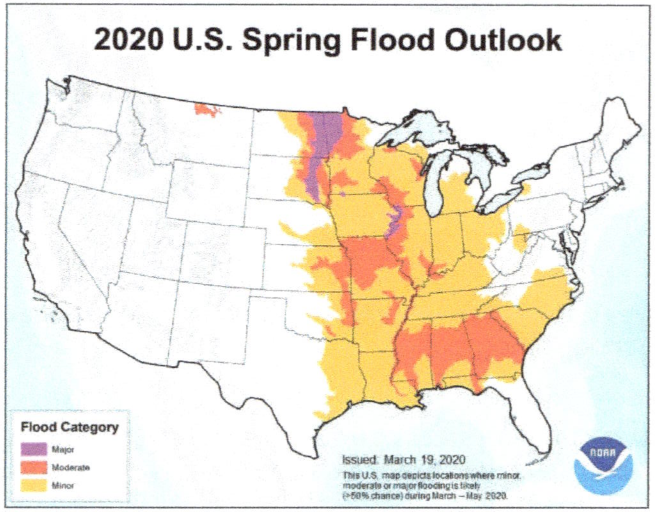

An example of a flood map. NOAA

What technologies and design solutions exist to detect, warn, or reduce damage? Floods

Forecast and Detect

- Rainfall amounts are constantly tracked.
- River levels are constantly tracked.
- Dams are inspected for weakness.
- Duration of potential storms that could cause flooding are monitored.
- River drainage basins are monitored for leaking or breakage.
- The soil moisture is monitored to see if the soil can absorb more rainfall or water.
- Ground temperature and snowpack are monitored.

Warn and Communicate

Flood alerts are issued in a number of ways:

- Weather radios
- Phone apps
- TV
- Radio
- Newspaper (if there is time)
- Local emergency management patrol areas to help alert and evacuate families, if necessary.

Reduce Damage

- Dams can release water slowly so that downstream communities are flooded less.
- If reservoirs are full when a flood happens, the reservoir water can be added to the aquifer and the new flood waters can go into reservoirs.
- Wetlands slow the flow of floodwaters and let sediments settle, and also, microorganisms in wetlands can filter pollutants from the water.
- Barriers can be built to prevent floodwaters from entering an area and can divert water to specific areas.

About Earthquakes

An earthquake happens when two plates slip, slide, or crumble as they move past each other. About 90% of earthquakes occur on plate boundaries. There are places in the world with more active plate movement, and these places tend to have more earthquake activity.

How much shaking is felt as the result of an earthquake is called the intensity. The intensity of an earthquake depends on the location where the earthquake started, the features of the plate(s), and the magnitude of the earthquake. Also, as you move further away from the earthquake's epicenter, its intensity decreases. Earthquake magnitude is measured using the Richter scale and is based near the epicenter of the earthquake.

- <u>Magnitude up to 2.9</u>: These earthquakes are so small that they are rarely felt by people and cause no damage to buildings.
- <u>Magnitude 3.0–3.9</u>: These earthquakes can be felt by people and may shake indoor objects slightly.
- <u>Magnitude 4.0–4.9</u>: These earthquakes shake and rattle indoor objects, and some objects may fall. Very little is felt outside.
- <u>Magnitude 5.0–5.9</u>: Everyone feels these earthquakes, and it can cause damage to buildings that are not well-built.
- <u>Magnitude 6.0–6.9</u>: These earthquakes have strong to violent shaking. Most buildings will be damaged, but buildings built for earthquakes will be only slightly damaged.
- <u>Magnitude 7.0–7.9</u>: These earthquakes cause damage to most buildings, and some may collapse. These can be felt from far away.
- <u>Magnitude 8.0–8.9</u>: These earthquakes cause major damage to buildings, destroying most average buildings. These earthquakes can damage large areas.
- <u>Magnitude 9.0 and above</u>: These earthquakes cause total destruction of large regions. Heavy damage extends far and can be felt hundreds of miles away. This magnitude tends to occur only once every 10–50 years.

Earthquakes happen suddenly and cannot be predicted. With larger earthquakes, several aftershocks can be expected within the first hour, and earthquakes of decreasing magnitude can last for days after the large earthquake.

Helpful Resources

CDC, Earthquakes: https://www.cdc.gov/disasters/earthquakes/index.html
Ready.gov, Earthquakes: https://www.ready.gov/earthquakes
Earthquake Drills: https://www.shakeout.org/dropcoverholdon/

Damaged buildings from an earthquake. Angelo Giordano

Car damaged from an earthquake. Marcello Migliosi

Students participating in an earthquake drill. Douglas Given/USGS

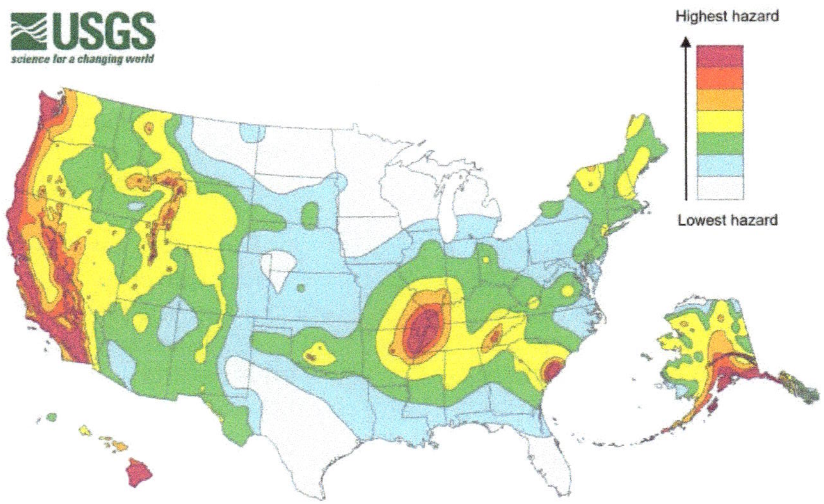

USA earthquake risk map. USGS

What technologies and design solutions exist to detect, warn, or reduce damage?

Forecast and Detect

- ShakeAlert: Warning systems exist all over the world. The United States Geological Survey has an early warning system called ShakeAlert. This system works by monitoring ground movement in different locations. The data is sent to monitoring stations, and warnings are sent out as quickly as possible.
- Smartphone apps: Certain smartphone apps can detect earthquake movement while an earthquake is happening. If an earthquake happens and this app is on a large group of user's phones, it can be used to sense the movement of the people and alert others farther away, potentially giving them time to get to safety.

Warn and Communicate

- Earthquake sensors can provide a warning of 2–10 seconds before the ground starts shaking in certain areas.
- ShakeAlert sends out warnings to computers and cell phones that include the expected magnitude and time before the earthquake begins in your location.
- Cities that normally experience earthquakes may have sirens to warn citizens of a coming earthquake.

Reduce Damage

- Buildings and homes can apply more earthquake-resistant design features, such as features like "shocks" or "seismic dampers" that can absorb the energy released in an earthquake.
- If living in an earthquake-prone area, make sure that all heavy shelves, pictures, and other items are secured.

Peer Feedback Guidelines

Giving Feedback to Peers

This tool was inspired by the Sticky Note Feedback resource originally developed by Ambitious Science Teaching at: https://ambitiousscienceteaching.org/sticky-note-student-feedback/.

Feedback needs to be specific and actionable.

For feedback to be productive, it needs to be related to science ideas and provide suggestions for improvement. Here are some examples of productive feedback:

- "Your model shows that the sound source changes position when it is hit. I think you should add detail about how the sound source moves back and forth after it is hit."
- "You said that the drum moves when it makes a sound, but the table does not move when it makes a sound. We disagree and suggest reviewing the observation data from the laser investigation."

Nonproductive feedback does not help other students improve. Here are some examples:

- "I like your drawing."
- "Your poster is really pretty."
- "I agree with everything you said."

How to give feedback

Your feedback should give ideas for specific changes or additions the person or group can make. Use the sentence starters below if you need help writing feedback.

- The poster said ____________________. We disagree because ______________. We think you should change ________________________________.
- I like how you __________________________. It would be more complete if you added __________________________.
- We agree that ______________________________. We think you should add more evidence from the ____________________ investigation.
- We agree/disagree with your claim that ___________________________. However, we do not think the _____________________ (evidence) you used matches your claim.

Receiving Feedback from Peers

The purpose of receiving feedback is to get ideas from your peers about things you might improve or change to make your work more clear, more accurate, or better supported by evidence. It can also help you communicate your ideas more effectively to others.

When you receive feedback, you should take these steps:

- Read it (or listen to it) carefully. Ask someone else to help you understand it, if necessary.
- Decide if you agree or disagree with the feedback, and say why you agree or disagree.
- Revise your work to address the feedback as needed.

Developing a Family Hazard Plan

Before, During, and After Checklist

During some natural hazards, the best plan is to shelter in place, which means you find the safest place in the house or building you are in. But, if necessary and if you have time, evacuation is often a safe response to an approaching natural hazard. Use this resource to learn about best practices for developing a family disaster plan, including how you might evacuate if necessary.

Below is a list of important considerations for you and your family. A comprehensive planning checklist is located here: https://www.ready.gov/plan .

BEFORE a disaster, plan for the following:	
• Learn about your local hazards and how to respond when they happen. Evacuation is NOT always the correct response.	• Sign up for emergency alerts, such as NOAA or FEMA alert systems. See https://www.ready.gov/alerts for a list of alert systems.
Plan how you will leave:	
• Identify the people who are safe for you to travel with (in case you are not at home).	• Know evacuation routes and identify alternate routes.
Plan where you will go:	
• Identify several destination options where you might go, such as the home of friends or family, or a safe motel or shelter where you can stay.	• Communities often identify local schools and community centers to serve as shelters for people who are evacuating.
Plan how you will meet with your family if you are separated:	
• Let out-of-area family members know where you are—they can help communicate for you. • Use social media to post that you are OK (but do not put your location for safety reasons).	• Call or text if cell service is good. • Contact local law enforcement or the American Red Cross, who can help reunite you with your family.
Plan what you will take:	
• Prepare a "go-bag" you can carry when you evacuate on foot or public transportation. Include supplies for traveling longer distances if you have a car. (See Kit List below.)	• Identify a safe place for family pets if you are unable to take them with you.

DURING a disaster:	
• Follow local evacuation instructions. Do not take short-cuts or go around roadblocks. • Take your emergency kit with you. Wear clothing that provides protection, such as long pants, a long-sleeved shirt, and a hat. • Take family pets with you if possible.	• Call contacts on your communication plan to let them know where you are going. Also, leave a note at your home stating where you are going. • Lock your home and unplug appliances (if time allows). • Check with neighbors to see if they need help evacuating.

Before traveling home:
If you evacuated due to a storm, check with local officials where you are staying and with those back home before you travel. Before you travel home, do the following: • Let friends and family know before you leave and when you arrive. • Fill up the gas tank in your car and gather supplies, such as water and food. • Charge all devices.

AFTER a disaster, be aware of the following:
• Always avoid power lines and appliances that might have been damaged in the disaster. • Listen to local officials about which neighborhoods and areas are safe for people and which places to avoid. • Reflect on what went well with your family's response and what you can do better next time, should a natural hazard occur.

Kit Checklist

The minimum you should have for an emergency kit are the following items:

- Water for 3 days
- Food for 3 days (non-perishable food is the type of food that does not need to be refrigerated)
- Flashlight and extra batteries
- First aid kit and any important medications
- Cell phone with charger
- Cash
- Any important documents or IDs (driver's license, student ID)
- Emergency contact and family information
- House keys
- Any special items for you or a family member

See https://www.ready.gov/kit for other kit items.

Family Communication Plan

When we evacuate, we will meet at:			
First Choice:		Second Choice:	
Our "out-of-area" emergency contact is:			
Name:		Phone #:	
Important phone numbers:			
Local Police Department/Precinct		Phone #:	
Local Fire Department/Station		Phone #:	
Other emergency contacts are (list name and best phone number):			
Parent:		Phone #:	
Parent:		Phone #:	
Sibling/Family:		Phone #:	
Sibling/Family:		Phone #:	
Other:		Phone #:	
Other:		Phone #:	

Our two evacuation routes are (name and sketch):	
Route 1:	Route 2:

Family Hazard Plan: Hazard-specific Information

Additional Wildfire Family Plan Information

Prepare

- Sign up for the local wildfire warning system.
- Know community risk, response, or evacuation plan.
- Find several ways to leave the area.
- Have a plan for livestock and pets.
- Gather emergency supplies, including an N95 respirator mask.
- Keep a fresh supply of everyone's medication.
- Designate a room that can be closed off from outside air.
- Set up portable air purifiers inside the house.
- Pay attention to air quality alerts.

Respond

- Listen to authorities to find out when to evacuate and when it is safe to return.
- Listen to EAS, NOAA Weather Radio, or local alert systems for current info and instructions.
- If evacuation is not required, stay inside where smoke levels are lower.
- Do not use anything flammable inside the house.
- If outside, do the following:
  - Look for shelter among rocks, lie face down, and cover your body with soil.
  - Breathe air close to the ground.
  - Avoid hot ash, charred trees, smoldering debris, and live embers.
  - Wear gloves and heavy-soled shoes to protect hands and feet.
  - Avoid fallen power lines.
- If trapped, call 911 and give them your location. Turn on lights to help rescuers find you.
- Reach out to family and friends using text, email, or social media. Make phone calls only in emergencies.

Additional Thunderstorm Family Plan Information

Prepare

- Know your area's risk for thunderstorms, tornadoes, lightning, and hail.
- Get a weather radio and know your local siren and alert sounds.
- Identify nearby, sturdy buildings close to where you live, work, study, and play. Practice going there.
- Cut down or trim trees that may be in danger of falling on your home.
- Consider buying surge protectors or a lightning protection system to protect your home, appliances, and electronic devices.
- Know the signs of a tornado: a rotating, funnel-shaped cloud; an approaching cloud of debris; or a loud roar—similar to a freight train.
- Watch for tornado danger signs: dark, often greenish clouds—a phenomenon caused by hail; wall cloud—an isolated lowering of the base of a thunderstorm; cloud of debris.
- Secure items that can be picked up by the wind. and bring companion animals indoors.

Respond

<u>If you are under a thunderstorm or tornado warning, find safe shelter right away!</u>

- When thunder roars, go indoors!
- For tornadoes, go to a safe room, basement, or storm cellar.
- If you are in a building with no basement, go to a small interior room on the lowest level. Use blankets to protect yourself.
- Stay away from windows, doors, and outside walls.
- Avoid running water or using landline phones.
- Unplug appliances and other electronic devices. Secure outside furniture.
- If boating or swimming, get to land and find a sturdy, grounded shelter or vehicle immediately.
- If necessary, take shelter in a car. Do not touch metal.
- Avoid flooded roads.
- Do NOT park your car under an overpass or bridge in a tornado. You're safer in a low, flat location.
- Use your arms to protect your head and neck.
- If in a mobile home, vehicle, or camper, find access to a sturdy shelter.

Additional Winter Storm Family Plan Information

Prepare

- Winterize vehicles, and put on winter tires or chains.
- Keep the gas tank in your vehicle full.
- Keep a supply of non-clumping cat litter to make walkways and steps less slippery.

Preparing your home:

- Protect pipes from freezing.
- Make sure windows are sealed.
- Consider buying emergency heating equipment.
- If you have a fireplace, keep a supply of firewood or coal.

Respond

- Stay indoors, wear warm clothes.
- Bring your animals inside.
- Make sure animals have liquid drinking water.
- Lower the thermostat, close off unused rooms, and cover windows at night.
- Knock ice and snow off of your roof using a roof rake.
- Clear snow and ice from furnace vents.
- Use ice melt and shovel walkways.
- Wear warm layers of clothing.
- Cover your mouth to protect your lungs from cold air and keep dry.
- Avoid doing too much work outside.
- Walk carefully like a penguin to help avoid slipping.

Additional Hurricane Family Plan Information

Prepare

- Consider flood insurance.
- Make a plan for your family and pets. Review the local evacuation plan.
- Know your nearest hurricane shelter location.
- Have flashlights, NOT candles.
- Purchase a sump pump and make sure it is working properly.
- Stockpile emergency protective items, like plywood, plastic sheeting, and sandbags.

Respond

- Make sure you have at least a half tank of gas.
- Do not drive on any water covered or flooded roads.
- Do not wade into floodwaters.
- Do not enter a building until it has been inspected for mold, gas leaks, live electric lines, and dangerous debris.
- Wear equipment (i.e., mask, gloves, boots) to protect yourself.
- Do not use electrical equipment around water.
- Do not drink tap water until the local authority says it is safe.

Additional Extreme Heat Family Plan Information

Prepare	Respond
• Know the location of local cooling centers, libraries, and other public places with air conditioning. • Make sure fans, air conditioners, and other items that keep people cool are working properly. • Plant trees to provide shade. • Add lighter colored surfaces or grass to your outdoor area, if possible. • Check that you have easy access to water.	• Wear light colors and loose fitting clothing. • Spend time in air conditioning at home, the local library, cooling center, or mall. • Try to keep out of the sun. • Drink a lot of water. • Avoid drinks with caffeine, sugar, or any other beverages that can dehydrate a person. • Eat cool foods such as salads. • Monitor yourself for signs of heat-related illnesses. • Monitor pets, kids, the elderly, and other people in your area to make sure they are staying cool. • Move any outdoor work that must happen to cooler parts of the day. • Monitor your electricity use and only use what you need. • Open up windows if the temperature has dropped and it is cooler outside than inside temperatures.

Additional Drought Family Plan Information

Prepare	Respond
• Use native plants in your yard. • Use a watering system for your plants or crops that requires less water or wastes less water. • Build a water storage system like a rain barrel. • Store water in ditches along fields. • Look at irrigation pipes and other areas for water leaks. • Raise animals that do not use as much water. • Farm native crops.	• Practice water conservation. • Only flush your toilet when needed. • Use a bucket to catch unused shower water to water plants. • Only use the dishwasher when it is full. • Allow lawns to die in extreme drought. • Check on those at risk to make sure they have access to drinking water.

Additional Flooding Family Plan Information

Prepare	Respond
• Move important items or documents to a high point or room. • Place important documents into waterproof containers. • Store pictures of important documents onto the cloud. • Clear the drains and gutters. • Check your basement for leaking. • Check your sump pump.	• Boil water during a flood. • Turn off propane tanks. • Unplug appliances and turn off utilities if instructed to do so. • Do not go through flooded areas. • If inside a building, go to its highest level. Only use the roof if absolutely necessary.

Additional Earthquake Family Plan Information

Prepare	Respond
• Practice drop, cover, and hold on. • Fix structural issues in your house. • Consider adding earthquake insurance. • Pack a disaster kit and make a meetup plan for after a disaster with your family. • Find a friend or family member out of state who can check in during an emergency. • Store heavy or large objects near ground. • Secure items that are on walls. • Avoid placing pictures or mirrors near where someone sleeps or sits. • Secure flammable products in a closed cabinet close to the floor. • Check gas and water lines. • Secure water heater.	**If in your home or another building:** • drop, cover, and hold on; • Do not remain in a doorway; and • move away from heavy or large objects that can fall. **If in a car:** • pull over to a clear location; • keep your seatbelt on; • watch for any falling rocks or debris; • avoid bridges, elevated roads, and powerlines; and • If a powerline falls on your vehicle, do not get out. **If outside:** • Find a clear spot on the ground and protect your head and neck. • Get as far away from buildings and tall or heavy objects.

Readings

What happened in Ryoishi Bay?

About the history of tsunamis at Ryoishi, Japan

Ryoishi is a small fishing village and bay located just inside the Iwate Prefecture in Japan. (A "prefecture" is a district.) The village is nestled into a valley at the edge of the ocean, surrounded by hills covered in pine forests. Many of the families in Ryoishi have lived there for many generations. Fishing is an important activity to Ryoishi. This makes access to the bay and the open ocean very important for its economy.

Over the years, Ryoishi Bay has been struck by several large tsunamis. In 1933, a large tsunami hit the area and wiped out many houses and businesses. After that, the government identified a "resettlement area" where they thought people could safely rebuild their homes. By 1948, a new seawall was in place to protect the village. New homes were then built in the coastal area again. The people of Ryoishi trusted that the seawall would reduce the damage from any future tsunamis. By 1980, the seawall in Ryoishi Bay was built even higher. As its height was increased, more people returned to live in the area behind the seawall to be closer to their livelihoods: fishing and coastal tourism. In 2011, about 600 people were living in Ryoishi.

Where is Ryoishi?

Ryoishi is in the northern part of Japan, on the eastern coast. Below is a series of maps that zoom into Ryoishi's location.

Zoomed out!

Zoomed all the way in!

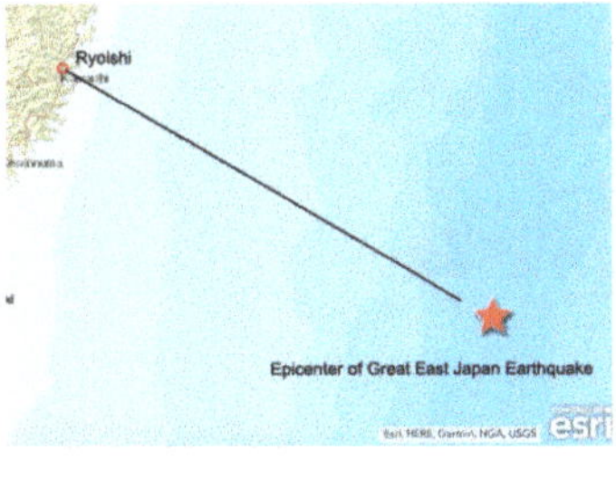

Where did people live in Ryoishi before the tsunami?

The following pictures show how Ryoishi grew after the 1933 tsunami, and how the housing changed from 1948 to 2010. The red line shows where floodwaters settled after the 1933 tsunami occurred. Once the seawall was constructed (shown by the dotted line) in 1977, people moved closer to the ocean. The people of Ryoishi assumed the area would be safe from future tsunami events.

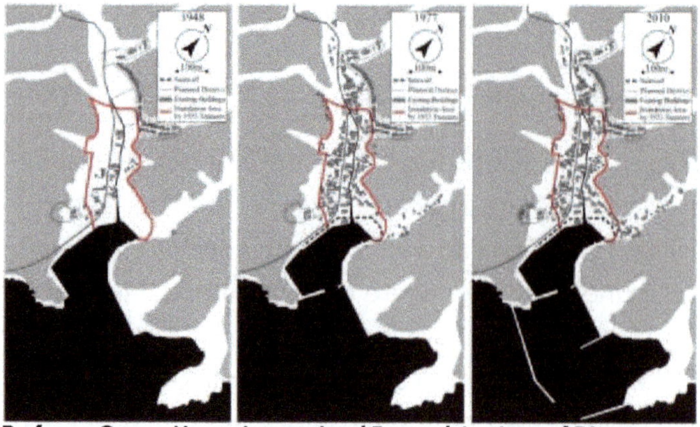

Professor Osamu Murao, International Research Institute of Disaster Science, Tohoku University.

Ryoishi and the 2011 tsunami

In 2011, the village was struck by another tsunami. The tsunami traveled at speeds of up to 500 miles per hour, but slowed down as it approached the shore. The seawall was hit by 60-65 foot waves, which went up and over the structure. The seawall broke under the weight and force of the water, causing water to rush into the community. Ninety-six percent of all the boats in the area were destroyed. Most homes in the red area (see the image above) were wiped out by the water. After the 2011 tsunami, people eventually started to build in the red area again (as seen in the images on slide C).

Close-up of the broken seawall after the 2011 tsunami. Ko Sasaki/The New York Times/Redux

Comparing Ryoishi to Nearby Communities

About Ryoishi:

In 2011, Ryoishi was a small village with an active fishing community and a population of about 600 people. Over half of the residents had family that had lived in the area for generations. The residents were mainly fishermen who wanted to live near their work on Ryoishi Bay. The larger commercial boats, however, were located farther south, near the main port of Kamaishi.

About the City of Kamaishi:

In 2011, the city of Kamaishi was home to over 63,000 people. It was once a big steel-producing city, but its main industries had switched to ecotourism and commercial fishing. To reduce the city's risk of tsunami damage, a series of breakwater designs was put in. Kamaishi was listed in the *Guiness Book of World Records* as having the deepest breakwater in the world, going 207 feet underground. Even though this impressive breakwater was meant to protect Kamaishi, the city still lost a third of its homes during the 2011 tsunami. Kamaishi's breakwater also suffered a lot of damage, but the breakwater design did reduce the average height of the waves to 29 feet. Unfortunately, it is believed that the breakwater made the waves heading into Ryoishi twice as large as the waves that hit Kamaishi. After the tsunami, the breakwater protecting Kamaishi was rebuilt.

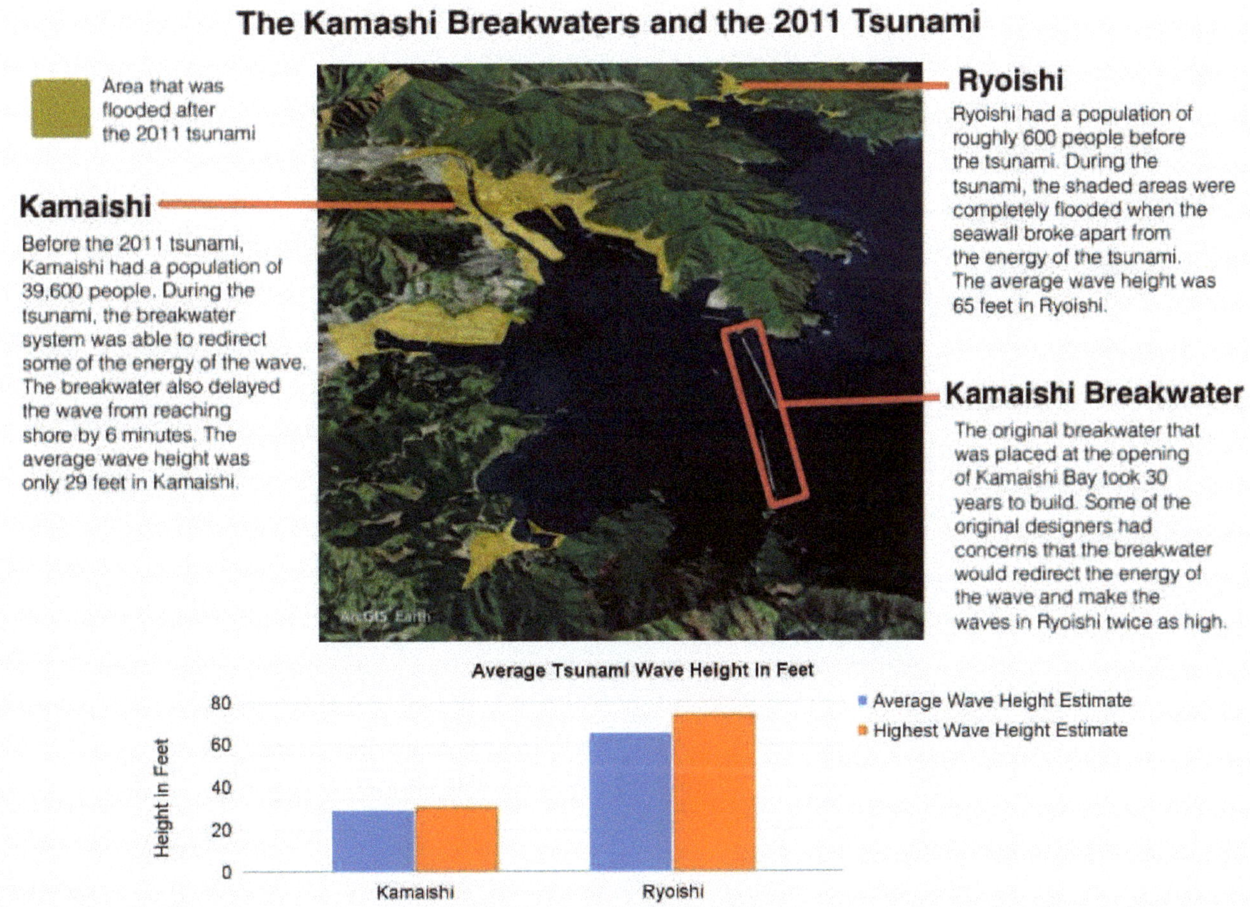

Source: Onishi N. (2011). Japan revives a sea barrier that failed to hold. New York Times. Retrieved from: https://www.nytimes.com/2011/11/03/world/asia/japan-revives-a-sea-barrier-that-failed-to-hold.html

How are tsunamis detected and warning signals sent?

Why did a tsunami reach Japan in 2011?

Earlier we read that a magnitude-9.0 earthquake happened underwater off the eastern coast of Japan on March 11, 2011. We figured out that earthquakes like this one are strong enough to cause large parts of the ocean floor to crack, tear, and move. These movements of the ocean floor push water away from the location of the earthquake (epicenter) in all directions. Very large waves form and can travel at speeds of 500 miles per hour. They can reach coastlines around the world in less than a day! In 2011, the first waves reached parts of Japan in only 8 minutes. With so little time to respond, detecting tsunamis quickly and accurately is essential.

How do we detect a tsunami?

As we read above, an underwater earthquake started a chain of events leading to the 2011 tsunami. So, this means that we must first detect the earthquake that causes a tsunami! Scientists on land have special instruments called seismometers that detect and record earthquake activity. By using seismometers in different locations around the world, we can figure out an earthquake's location, magnitude, and depth. This information is important to scientists. Remember we have learned that not every earthquake causes a tsunami. In fact, earthquakes must meet 3 criteria in order for a tsunami warning to be sent.

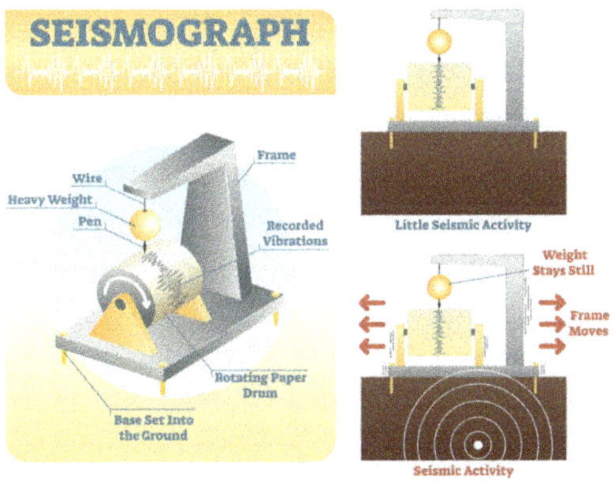

Think back to what we learned in previous lessons. What types of earthquakes cause tsunamis?

In order to cause a tsunami, an earthquake must meet 3 criteria. First, it must occur underwater. Most earthquakes that occur on land do not cause large movements in the ocean floor. That means they do not lead to water movement and waves. Second, an earthquake must release a lot of energy to cause large parts of the ocean floor to move suddenly. The amount of energy released by an earthquake is measured as its magnitude. Magnitude is sometimes referred to as the strength of an earthquake. Weaker earthquakes may not move the ocean floor enough to cause a large wave to form in the ocean. Lastly, an earthquake must be shallow and not deep inside Earth's crust. Shallow earthquakes tend to occur just below the ocean floor. This can cause the ocean floor to move more.

How do we locate a tsunami?

The second step of the tsunami warning system begins when seismometers detect an earthquake that might cause a tsunami. This step uses a system of special instruments. The instruments are located on the surface of the ocean, on the ocean floor, and even up in space! This system of instruments is called DART II, or Deep-ocean Assessment and Reporting of Tsunamis. When the seismometers on land detect an earthquake that might cause a tsunami, they send a signal to a satellite that is orbiting Earth. The satellite then sends signals to several floating buoys on the surface of the ocean near the earthquake's epicenter.

Next, each surface buoy sends a signal to wake up an instrument down on the ocean floor called a tsunameter. You might be wondering why we don't send the signal from the satellite directly to the tsunameter--it turns out that satellite signals do not travel well through water. Fortunately, sound travels very quickly through water. So, sound waves, or sonar, are used to send and receive signals from the tsunameter.

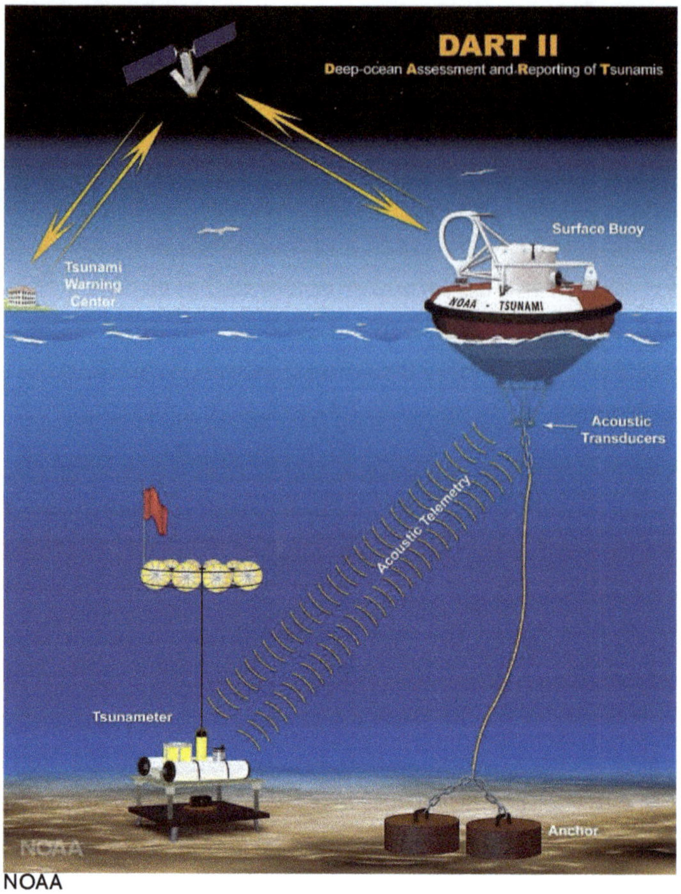

NOAA

How does a tsunameter work?

A tsunameter is anchored to the ocean floor. It has several devices attached to it that float upward. As a wave moves across the ocean's surface, the water underneath the wave gets pushed down deeper into the ocean. If a large wave occurs, more water is pushed down, which also pushes down on the floating devices attached to the tsunameter. The tsunameter records the height and movement of these floating devices every 15 minutes when it is in "sleep" mode. When an earthquake occurs nearby and the tsunameter is switched to "active" mode, it takes measurements every minute.

Stop to Wonder: Why do you think the tsunameter is sometimes in sleep mode and not always in active mode?

When a tsunami occurs, a huge wave moves across the ocean's surface, pushing large amounts of water down on the devices attached to the tsunameter. Once the devices sink to a certain depth, the tsunameter sends a signal back up to the surface buoy telling it that a tsunami was just detected on the ocean surface.

When is a tsunami warning sent?

On the ocean surface, the surface buoy sends a signal to the satellite. The satellite then sends a final message back to the scientists on land. The scientists use computers to predict whether the tsunami will reach the shore. If they think the tsunami will reach shore, then they work to figure out the answers to these questions: (1) how long will it be until the tsunami hits, and (2) how tall will the wave be when it hits? The answers to these questions can be used to help save lives.

Sources:

- US National Oceanic and Atmospheric Association (NOAA). Retrieved from: https://www.noaa.gov/explainers/us-tsunami-warning-system
- Bureau of Meteorology, Australia. Retrieved from: http://www.bom.gov.au/tsunami/about/detection_buoys.shtml

Voices from Japan Tsunami Survivors

Account 1: Kikuchi Nodoka, third-year student, Kamaishi Junior High School	Account 2: Sho Komine and Yasushi Kaneko, staff writers for the Yomiuri Shimbun
"I thought, *The tsunami will come soon after the shaking stops. I must evacuate to the surrounding hills as quickly as I can.* As soon as this idea entered my mind, my body reacted, thanks to the routine emergency drills that I had experienced before." Kikuchi and other students, including elementary students in the school next door, evacuated immediately and went to higher ground in the hills nearby.[1]	When the tsunami reached Kamaishi, the floodwaters reached the third floor of one local junior high and an elementary school. The schools' announcement system malfunctioned during the earthquake, so the teachers were not able to announce for students to evacuate. However, both schools had been practicing drills for disaster situations and knew what to do. The junior high students helped the younger elementary students as they ran to higher ground. One student, Dai Dote (age 13), helped two elementary students and kept them calm as they climbed a nearby hill. Dote told them it would be OK and made sure they were safe.[4]
Account 3: Hideko and Waichi Nagano	**Account 4: Yu Muroga, medical salesman**
Hideko and Waichi Nagano watched from the hills above their town as water swept through the entire area. "It was a huge black mountain of water, which came on all at once and destroyed the houses," he said. "It was like a solid thing."[8]	"When the earthquake hit, I had left my house and was driving. So I wasn't sure if it was an earthquake or just my car. But when I saw the way people were moving and the buildings shaking, I knew it was an earthquake." "I'm originally from an inland area, so I didn't have the common sense to know there was a danger of a tsunami when an earthquake happens."[5]
Account 5: Hiroko Sugawara, grandmother	**Account 6: Yuichi Owada, volunteer firefighter**
"We wake up and see the sea, we talk about the sea ... whether it is quiet or if it has waves, as we have always done together. When you live by the sea, you depend on it, and it is part of your life..." "When we saw the water approaching, it flooded the national road as two trucks were passing. And then the water just kept coming."[5]	"My house was located just in front of the sea. I lived there for forty years. The sea was like our great mother. It was like our treasure. We have always received blessings from the sea..." "We had a levee that was 5 meters (15 feet) tall that we thought could protect against a tsunami. And so we based all of our evacuation areas and exercises on this assumption."[5]

Account 7: Nobuko Kikuchi, 72 year-old resident of Iwanuma, Miyagi Prefecture	Account 8: Kenji Saito, business owner
She "couldn't hear the emergency sirens that followed the 9.0 magnitude earthquake that struck on March 11, 2011. Nor could she hear the public announcement urging people to evacuate to higher ground as a massive tsunami was approaching the coast of northeastern Japan's Tohoku region. Kikuchi is deaf. She owes her life to a neighbor who came to alert her. Kikuchi narrowly escaped the monster wave, which uprooted and washed away her house." [7]	"All of a sudden, the room started shaking. I grabbed my computer so it wouldn't fall, and I waited for the shaking to stop. I thought for sure a tsunami was going to come. I started yelling at people in the office to run to safety as soon as possible..." "Then the water started rising, but I still thought it would be OK because of the levee. But as time passed, the water rose above the levee ... As the water got higher, houses were swept up and washed away." [5]
Account 9: Ayumi Osuga, mother with three small children	**Account 10: Masao Sato, builder and florist**
Ayumi Osuga was in her room practicing origami with her three young children—all between the ages of 2 and 6. Osuga ran to meet her husband, packed up her kids in the car, and sped away to a hilltop where her family had a home. "My family, my children. We are lucky to be alive," she said upon returning to her home. "I have come to realize what is important in life." [2]	I was worried about my home. So, I watched from some stone stairs nearby. It was strange, as I looked towards the sea, I couldn't see the shape of the waves. As the houses and buildings were being destroyed, dust flew into the air. All you could see was dust....Then I saw my house and car washed away." [5]
Account 11: Kuniko Suzuki, grandmother	**Account 12: Masako Sugawara, mother and fish factory worker**
Even though we saw the car passing by to warn us, we didn't actually think the tsunami could reach us because we lived quite far from the ocean and we were on the hill. [5]	I thought we were in danger, so I yelled to my husband, we have to go to higher ground. I saw electric wires being cut and falling down. It was scary and I shouted that everyone needed to come higher. [5]
Account 13: Yasuo Kishi, retired fisherman	**Account 14: Jessica Besecker, an American school teacher in Japan**
There was too much water, and I couldn't get to the steel buildings, so I just climbed up a tree. [5]	The tsunami alarm went off. The announcements were in Japanese, and [I] couldn't understand them. [3]

Account 15: Katie Oi, US teacher at a Japanese middle school
"When (the earthquake) finally ended, we rushed outside gathering the students in front of the school. Students were panicked and many were crying. We were instructed to sit down as we waited for the aftershocks while teachers ran back inside to collect blankets, the students' jackets, and bags. The weather forecasted snow and for good reason: it was cold... I had no idea that the tsunami alarms were ringing, warning the town to evacuate immediately to higher grounds. However, the elementary school students from below arrived within ten minutes, and townspeople had gathered in our school parking lot... As we continued to wait, we heard a loud crash in the distance that many mistook for thunder, when in reality it was the main highway bridge destroyed by the tsunami's force ... We had to move to higher ground." [6]

These tsunami survivor stories are quoted from the following sources:

1. Akira, U. (2019). *Preparing future generations for another disaster.* Government of Japan, Public Relations Office. Retrieved from: https://www.gov-online.go.jp/eng/publicity/book/hlj/html/201912/201912_04_en.html

2. Chappell, B. (2011). For Japan's quake survivors, 'It's urgent everywhere.' *NPR.* Retrieved from: https://www.npr.org/sections/thetwo-way/2011/03/13/134510288/in-japans-north-evacuees-face-radiation-fears-uncertainty

3. Flock, E. (2011). Japan's earthquake and tsunami: An American teacher's account. *The Washington Post.* Retrieved from: https://www.washingtonpost.com/national/japans-earthquake-and-tsunami-an-american-teachers-account/2011/03/29/AFH9m6xB_story.html

4. Komine, S. & Kaneko, Y. (2011). Local wisdom a lifesaver for kids. *Daily Yomiuri Online.* Retrieved from: http://www.ce.gunma-u.ac.jp/news/wb_20110329.pdf

5. Nicholson, P. (Director). (2011). *Japan's Tsunami: Caught on camera* (documentary). UK: Channel 4 News. Retrieved from: https://www.youtube.com/watch?v=oArd_9uZOnE

6. Oi, K. (2011). *Firsthand account of Japanese earthquake and tsunami from Katie Oi '10.* Claremont McKenna College. Retrieved from: https://www.cmc.edu/news/firsthand-account-of-japanese-earthquake-and-tsunami-from-katie-oi-10

7. Otake, T. (2013). Filmmaker captures the 3/11 stress of Tohoku's deaf. *The Japan Times.* Retrieved from: https://www.japantimes.co.jp/culture/2013/03/10/films/filmmaker-captures-the-311-stress-of-tohokus-deaf/

8. Perry, R.L. (2017). The school beneath the wave: The unimaginable tragedy of Japan's tsunami. *The Guardian.* Retrieved from: https://www.theguardian.com/world/2017/aug/24/the-school-beneath-the-wave-the-unimaginable-tragedy-of-japans-tsunami

Case Study: Kamaishi East Junior High School

The students of Kamaishi

On March 11, 2011, the Great East Japan Earthquake occurred at 2:46 pm. It was a sunny Friday afternoon.

The earthquake had a magnitude of 9.0. The earthquake generated a tsunami that traveled quickly to the east coast of Japan. Tsunami warning signals were sent within minutes. The tsunami arrived on the shores of coastal communities within 10–30 minutes after the earthquake.

The Miyagi and Iwate regions of Japan were some of the hardest hit coastal communities. The town of Kamaishi is within this part of the coast and was devastated by the tsunami. Over 1,000 people died in Kamaishi that day. That number included 5 school-aged children who were not in school the day of the tsunami. But 99.8% of all schoolchildren enrolled in the junior high and elementary school survived.

USAID/USGS

When the tsunami warning was sent to Kamaishi that day, students at Kamaishi East Junior High School knew what to do. They evacuated for higher ground immediately, helping both younger students from the neighboring elementary school and elderly residents in the homes nearby. The almost 300 junior high school students are responsible for saving almost 3,000 lives that day.

What did the junior high students do?

Most of the schools in Japan practiced earthquake and tsunami drills, which often consisted of protecting oneself during the earthquake, and then evacuating to the playground, stadiums, or upper floors of the schools in response to the tsunami.

During the 2011 earthquake, the students knew the earthquake was a big one and would cause a tsunami. The earthquake caused the junior high school alarm system to break before tsunami warnings were sent, but the students knew what to do. They knew to evacuate to higher ground in the hills near the school, and to do so quickly. As they evacuated, the students did the following:

- They didn't hesitate to evacuate because they knew the way to respond in the situation.
- They stayed calm and headed to an evacuation zone.
- They shouted, "A tsunami is coming!" to warn nearby residents.
- They helped elementary school children and elderly residents.

How did they know what to do?

For several years the students at the junior high school had been working with an engineer, Toshitaka Katada, to learn how to respond during a natural hazard. Katada spent 10 hours each year helping students plan their response for when an earthquake or tsunami might happen.

Toshitaka Katada

It is through this education that the students knew how to respond and where to go quickly to save their lives and the lives of other students and community residents.

Consider these questions:

- Have you had a similar experience where you have practiced what to do during a natural hazard in your area?
- Do you know how to respond?
- What questions would you have and do you know how to find the answers?

Sources

- Government of Japan, Public Relations Office. (2013). *Lessons from disasters*. Retrieved: https://www.gov-online.go.jp/eng/publicity/book/hlj/html/201303/201303_02.html
- Government of Japan, Public Relations Office. (2013). The 'Miracle of Kamaishi': How 3,000 students survived March 11. *JFS Newsletter*, 133.
- Kamiya, S. (2011). Students credit survival to disaster-preparedness drills. *The Japan Times*.
- No author. (n.d.). *Disaster management education in Kamaishi*. Retrieved: https://www.chicago.us.emb-japan.go.jp/7stories4.pdf

Sending Warning Signals

How we communicate with one another, including how we warn others in an emergency, looks very different today than in the past. As you read this text, consider how technologies have changed over time to make communication between people more reliable, faster, and overall easier.

The History of Sending Warning Signals

If you lived a very long time ago, the only way you could warn others about danger would have been to use something that another person could detect directly using one or more of their senses. To alert them using their sense of hearing, for example, you might have yelled or hit something like a hollow tree trunk or a drum. To alert them using their sense of sight, you might have created a fire with a bright flame or a lot of smoke.

Great Wall of China William Olivieri

In the past, humans set up chains of communication where warnings could be passed from one person to the next over longer distances faster than someone could run in a short period of time. For example, evidence suggests that, up to 4,000 years ago, soldiers along the Great Wall of China lit fires to send smoke signals from tower to tower to warn each other if they saw an enemy army approaching.

The Telegraph. During the 1800s, people started sending warning messages over much longer distances using signals generated by electricity. The telegraph is an example of using signals generated by electricity to send messages. A telegraph used a series of long and short taps called Morse code to send a signal.

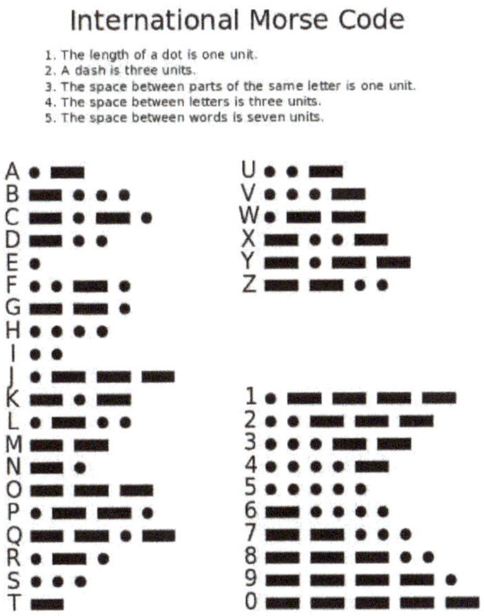

Morse code. Rhey T. Snodgrass & Victor F. Camp, 1922

A telegraph message could be sent over very long distances very quickly, but this method of communication required someone trained to send the signal in Morse code and also someone to be able to read Morse code to receive it. Another disadvantage is that telegraphs were mostly sent by wire, and the signal could only be sent through a wire to another telegraph location.

After a warning message was received by telegraph, it then had to be delivered by hand or verbally from one person to another. The people that needed the message could be miles away and a lot of messages were delivered by people traveling from place to place, or by using bells or sirens.

The Telephone. In the late 1800s and early 1900s, the use of telephones replaced telegraphs because telephones allowed people to communicate by voice, without relying on the codes of telegraphs. As telephone technology improved, the telephone wires were brought all the way to people's homes and businesses.

Telephone switch operators. US National Archives

At first, people had to connect the many different phone lines that were wires by hand, but soon technology was made to allow any telephone to be connected to any other telephone by a system of switches.

The image shows phone line operators connecting wires by hand. Each phone had to be called separately, making it a slow way to warn people of any hazard, and only some households had phones. Even today, most people can only place one phone call at a time.

Loudspeakers and Sirens. Another way that people are alerted to warnings is by loudspeaker. The speaker could be much louder and be heard further away than someone who is only yelling on the street, and could be used by someone who just received a warning signal.

Loud sirens became a common way to alert people during an emergency, and sirens are still used today. An example of this is the tornado siren. Unfortunately, sirens like a tornado siren are mostly designed to alert people who are outside of their houses, and can be hard to hear inside buildings.

The Radio. After the development of the telegraph, scientists and engineers designed radios. Radios send signals and messages wirelessly across an area. We still use radios in our homes, schools, businesses, and cars today. One advantage of the radio is the ability to send messages out to a large area, and not just to one location using a wire.

Rod Flores

Stop and Think: How have emergency communication systems changed over time?

Modern Warning Signals Sent by Radio

The most reliable system we have to send messages to large groups of people is by radio. Unfortunately, traditional radio signals can be unreliable because of old equipment and other signals interrupting the message. Also, in order to hear an alert on the radio, the radio has to be turned on all the time.

Consider listening to a radio station. If you are listening near the source of the station, like the radio tower, the signal comes in clearly to your car or radio. But as you move further away from the station, the signal can become weak or have more static. As the signal becomes weaker, it is called a degraded radio signal. Radio signals become degraded, or reduced, as they travel over long distances. This happens partly because the signal becomes weaker as it travels away from its source. This is like when light gets dimmer further away from the light source and tsunami waves get smaller as they travel away from their source. The radio signal is carried by energy and the energy spreads out more and more as it travels. This means there is less energy traveling through any location farther from the source than there is closer to the source.

The problem of radio signals getting weaker as they travel can be solved by adding devices called repeaters. A repeater receives a weak signal and then retransmits it as a stronger signal. Unfortunately, these devices can be expensive to build and maintain in remote locations.

Another solution is to encode the radio signal in a digital format. This solution can be much less expensive than installing repeaters.

Signals encoded in a digital format can be transmitted over longer distances than signals that are sent using the analog format that radios have traditionally used. Like analog signals, digital radio signals get weaker as they travel. However, digital radio receivers can translate much weaker signals back into their original format better than analog radio receivers can.

How Digital Signals Transmit Messages

Let's think about how digital signals, like a tsunami warning over a loudspeaker, start. First, a person speaks into a microphone. The microphone converts the sound of the person's voice into an electrical signal that can be transmitted along a wire. (You will learn how microphones convert sounds into electrical signals and how speakers convert them back into sounds in the *Magnets Unit*.) The electrical signal can be displayed as a graph.

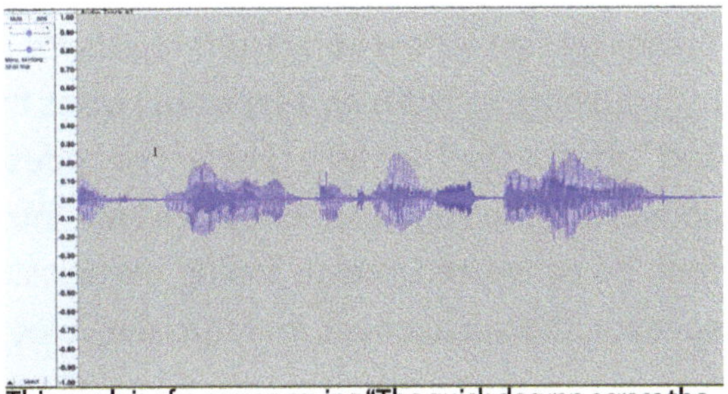
This graph is of a person saying "The quick dog ran across the yard."

This analog electrical signal can be converted to a digital signal by recording the value, or location, of the signal at regular intervals. These values are called samples of the original analog signal. Sometimes the signal we see generated shows the values sampling of the audio on the Y-axis. The image below shows an analog signal in orange and also shows the times when data in the signal is turned into a digital signal (the sample times, shown by the lines on the graph).

Stop and Think: The blue dots show the sampling. Look at the blue dots on both the analog signal and the digital sampling. How do they compare?

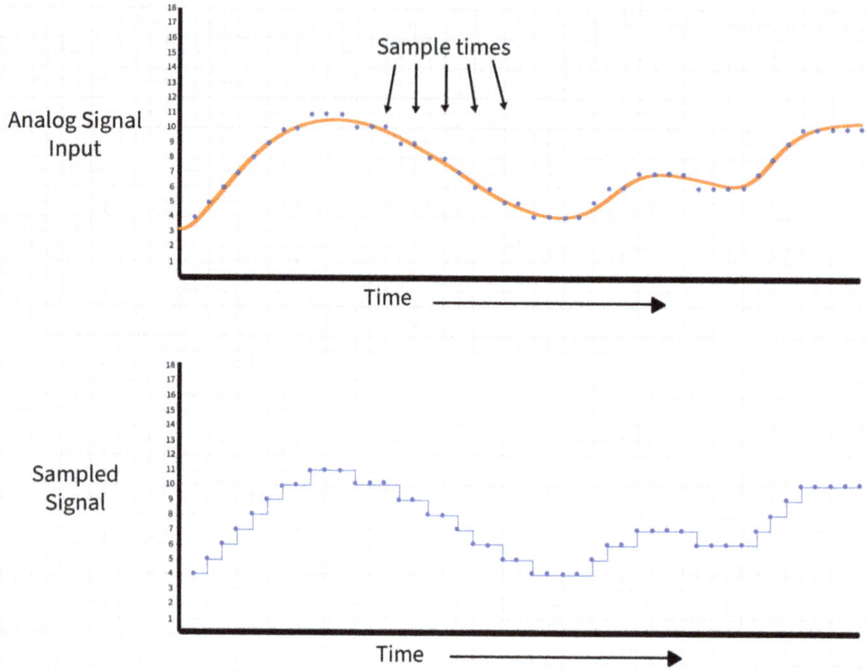

The sampling that happened above is the first step of analog encoding. The second step is changing the sample values from the graph above into a sequence of 1s and 0s (similar to Morse Code) that encode the digital output. For example, if the digital output section that was sampled above was at a 2, it would be encoded in the signal to 000010. See the image to see how the sound wave that was recorded and sampled above is turned into a digital signal.

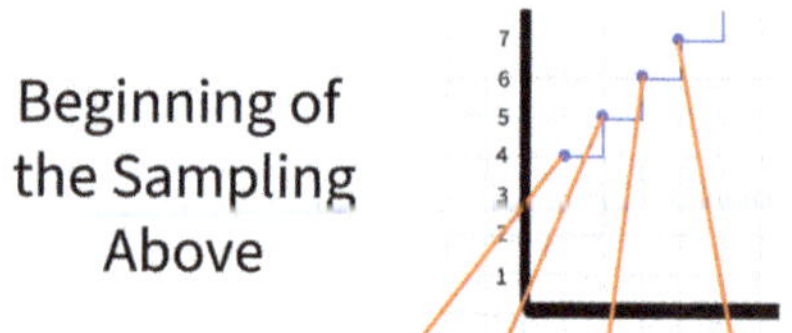

Sampled Values	4	5	6	7
Sampled Values in Binary Code	000100	000101	000110	000111

An analog signal, now converted to a digital signal, can then be sent by a radio transmitter as a series of pulses that are either "on" for 1s and "off" for 0s. The radio receiver decodes the digital signal by converting the sequence of pulses back into sampled values (like turning 000010 back into 6) which can then be converted into an analog, electrical signal. When this analog, electrical signal is connected to a speaker, the speaker will recreate the voice spoken into the original microphone.

The Benefit of Digital Signals for Transmitting Messages over Long Distances

Think about what happens when an analog signal, like a hazard warning signal sent over a radio, has to travel a long distance. The further the signal goes, the weaker the signal gets.

Now let's think about a digital signal. In this case, the size of the pulses on the signal is smaller, but the receiver only has to be sensitive enough to distinguish between something that is a 1 or a zero, not the difference between a .6 and a .5 or a .7. Because of this, the digital signal has a clearer sound further away than an analog signal. Once the 0s and 1s cannot be interpreted by the radio, the signal is no longer played through the radio.

Cell Phones and Digital Push Notifications

Today we have systems, like apps and push notifications, that allow us to warn thousands of people at once when a hazard, like a tsunami, is occurring. These digital signal notifications have helped to save thousands of lives during different emergencies by quickly alerting large groups of different people at once. The signal can also be translated by the phone system into the language the user is the most comfortable using. But, getting hazard warnings to people in some remote locations is still a challenge. Consider cell phone coverage maps from any cell phone provider. Do you ever notice that there are some areas that are not covered? While many people have cell phones, there are still remote places that do not have reliable telephone or cell phone service.

A tsunami alert sent out in Orange County CA after the 2022 Hunga Tonga-Hunga Ha'apai eruption

Tsunamis can occur in many locations, and affect many different people. We also know that many of those people may live in remote areas.

- When considering our stakeholder groups, what type of warning signals might make the most sense to alert them of a hazard?
- What would be a more reliable signal for them to receive, digital or analog? Why?
- Is a warning signal, like a radio signal or a loudspeaker broadcast, the only signal needed to warn people?
- If yes, why? If no, what other method(s) would we want to consider for all of the stakeholder groups we have identified?

www.ingramcontent.com/pod-product-compliance
Ingram Content Group UK Ltd.
Pitfield, Milton Keynes, MK11 3LW, UK
UKHW051307061126
16386UKWH00011B/48

9 781792 483189